RFID Essentials, Innovations and Beyond

RFID Essentials, Innovations and Beyond

Shiva Sukula

Ess Ess Publications
New Delhi

RFID Essentials, Innovations and Beyond

ISBN: 978-81-7000-651-0

Rs.975/-

First Published 2011

Published by:
Ess Ess Publications
4837/24, Ansari Road,
Darya Ganj,
New Delhi-110 002.
INDIA
Phones: 23260807, 41563444
Fax: 41563334
E-mail: info@essessreference.com
www.essessreference.com

Cover Design by *Patch Creative Unit*

Printed and bound in India at Salasar Imaging Systems.

CONTENTS

Foreword

The development and growth of technology in the area of RFID (Radio Frequency Identification) and its applications has been explosive, mainly due to its better adaptability and capacity compared to the Barcode Technology. RFID technology provides a combination of automation and security activities in the proper maintenance of items either within the organization or when the item goes out-of its premises. While Barcodes have become very popular in the retail trade for on the shelf items and other similar applications, they have limited capacity to provide large volume of data about the product they are applied to.

By using wireless radio communications to uniquely identify objects or people, RFID is one of the fastest growing automatic data collection (ADC) technologies. The RFID varieties are basically of passive and active type and these have their respective domains of applications. The passive technology comprises of one or more reader/interrogators and RF transponders in which data transfer is achieved by means of suitably modulated inductive or radiating electromagnetic carriers. The active RFID systems carries data in suitable transponders, generally known as tags, and retrieval of data is by machine-readable means, at a suitable time and place to satisfy particular application needs. These

RFID's use a combination of radio-frequency and microchip to provide the response of product ID. RFI chips are of particular interest, because they have become smaller and smarter to the point where they can be added to every kind of item including humans and can be read and updated from a distance. The adoption of RFID technology promises a solution that could make it possible to inventory hundreds of thousands of items in their collections in days instead of months.

The book by Shiva Shukla has a very wide canvass that talks of the RFID technologies and their applications in different areas. However, the RFID application in the Library environment and cataloging are her main focus. The book explores the RFID concept to make it widely understood or installed in the professional environment, and its application, standardization, and innovation. The chapters are designed to inform the readers about RFID's in the constantly changing technological arena.

Since adoption of this technology is still relatively new hence the book concentrates on discussing many features of the technology that are not well known and even little understood by the general populace..

The book includes chapters on various topics such as RFID System Elements, Tools, Techniques and Technologies behind RFID, Implementation Issues of RFID, RFID and Bar Code Technology, RFID Standards, RFID systems and Solutions for Library, RFID Applications in Various Arena, RFID and Consumer Privacy Issues, Near Field Communications and RFID in Indian and Global Perspectives. The book is thus a comprehensive source of information. While the book is very detailed and is inclusive of myriad aspects of

RFID but the matter has been delivered in such a way that it is suitable for layman as well as professionals.

I congratulate Dr. (Ms) Shiva Sukula for bringing out such as excellent work.

Prof. S. K. Kak
Vice Chancellor,
Mahamaya Technical University,
NOIDA.

PREFACE

Radio Frequency Identification (RFID) is a generic term used to describe technologies that involve the use of data stored on small chips or tags which can be communicated to a reader from a distance by means of radio transmission. It is consisted of the RFID tags themselves (which consist of an antenna attached to a microchip), the RFID readers, and the supporting database infrastructure (hardware and software). The technical capabilities and distinctions among RF technologies are evolving over time. RFID is an enabling technology with the unique advantage that it eliminates the requirement of human intervention thus providing immense business value. There is more to RFID than just tags that exchange data with readers wirelessly. It has tags and data structures that software applications can use. Few years ago, some companies realized the potential of this technology that it can make any object self-identifying to the information system. It was applied to track tools, and to trace various kinds of products. This technology can be used to mistake-proof mixed-flow assembly of products as PCs.

It is a fast moving technology, with evolving and competing standards from ISO, from the US-based EPCglobal consortium, and from individual companies. Further moving, it as found that the combination of RFID with kanbans offers opportunities for smoother integration with the overall information system. RFID can be the basis for effective and efficient access control in factories handling dangerous or sensitive materials. RFID technology may be hi-tech. It takes some management tricks to implement RFID technology applications to a particular situation or organization. It is necessary to know precisely about its strengths and weaknesses. Every technology comes with pros and cons.

Comparing RFID with bar codes, not only is the amount of data that can be stored in a bar code is much smaller than in an RFID tag but it cannot be updated. Reading barcodes requires human intervention, a clean, high-contrast environment, and often more than one attempt. RFID technology can be used to track tools, equipment or inventory, to monitor access to facilities or secure areas, or to monitor patterns of activity. Various RFID case studies are providing the information of successful application of RFID technology to variety of organizations as well as processes. There are already a variety of actual and contemplated uses of RFID. RFID systems can also be designed to enhance security and safety. While some uses of RFID may offer benefits to employees, RFID in the workplace also raises important privacy concerns also.

It is essential to address privacy issues inherent in RFID technology in advance of deployment. It includes determining whether it is appropriate or necessary to use RFID technology and incorporate privacy features in the emerging technology. It empowers and fosters the development of good practices before privacy invasive ones become entrenched. Using a standard identifier format makes it easier for organizations to decode identifiers. RFID represents a technological advancement in Automatic

Identification and Data Capture (AIDC) because it offers advantages that are not available in other AIDC systems. It offers these advantages because it relies on radio frequencies to transmit information rather than light, which is required for optical AIDC technologies. The ability of RFID technology to communicate without optical line of sight and over greater distances reduces the need for human involvement in the identification process. Today RFID technology makes the label dynamic by enabling it to acquire data about the object.

- **Shiva Sukula**

1
Radio-Frequency Identification

WHAT IS RFID?

Radio Frequency Identification, RFID is a dedicated short range communication (DSRC) technology. This term (RFID) is used to describe technologies using radio waves to identify people or objects automatically. RFID technology seems similar to the bar code identification systems; however one big difference between the two is that RFID does not rely on the line-of-sight reading that bar code scanning requires to work. In RFID, the electromagnetic or electrostatic coupling in the RF (radio frequency) portion of the electromagnetic spectrum is used to transmit signals. An RFID system consists of an antenna and a transceiver, which read the radio frequency and transfers the information to a processing device (reader) and a transponder, or RF tag, which contains the RF circuitry and information to be transmitted. The antenna provides the means for the integrated circuit to transmit its information to the reader that converts the radio waves reflected back from the RFID tag into digital information. This information is passed on to computers that can analyze the data.

RFID is the use of an object (typically referred to as an RFID tag) applied to or incorporated into a product, animal, or person for the purpose of identification and tracking using radio waves. Some tags can be read from several meters away and beyond the line of sight of the reader. RFID comprises *interrogators* (also known as *readers*), and *tags* (also known as *labels*). RFID is a wireless technology used to transmit information from tags attached to objects in order to automatically identify and track the objects. RFID is part of a broad category of automatic identification (Auto-ID) technologies. Auto-ID technologies include RFID, barcodes, optical character readers and some biometric technologies, such as retinal scans. Auto-ID technologies enable machines to scan, capture and "recognize" identification for an object. Machine-readable information (information that can be read by computers, scanners, etc) can be used to automate processes and to reduce errors. This technique reduces the amount of time and labor needed for humans to recognize and manually input data about an object. Some auto-ID technologies (barcode systems) often require a person to manually scan a label or tag to capture the data. RFID is designed to enable machines called readers to capture data on tags and transmit it to a computer system—without the need for a person to be involved.

RFID tags consist of a microchip attached to a radio antenna mounted on a substrate. The antenna is typically attached to a microchip. All RFID tags do not use silicon chips. The chip can store data. Some chips store only a unique numeric identifier. Chips may store data about an object (an asset, a product or a shipment) - for example, the date of manufacture, destination and product information, all of which may be written to a tag. Retrieving the data stored on an RFID tag requires a reading device called a reader or an interrogator. A typical reader is an electronic device that has one or more antennas that emit radio waves and receive signals back from the tag.

In RFID systems, the tags hold the data are broken down into two different types. Passive tags use the radio frequency from the reader to transmit their signal. Passive tags will generally have their data permanently burned into the tag when it is made, although some can be rewritten. Active tags are much more sophisticated and have on-board battery for power to transmit their data signal over a greater distance and power random access memory (RAM) giving them the ability to store up to 32,000 bytes of data.

AUTOMATIC IDENTIFICATION

Due to the procedures used for the transfer of power and data, contactless ID systems are called *RFID systems* (Radio Frequency Identification). In recent years automatic identification procedures (Auto ID) have become very popular in many service industries, purchasing and distribution logistics, industry, manufacturing companies and material flow systems. Automatic identification procedures exist to provide information about people, animals, goods and products. The omnipresent barcode labels that triggered a revolution in identification systems some considerable time ago, are being found to be inadequate in an increasing number of cases. Barcodes may be extremely cheap they have low storage capacity and cannot be reprogrammed. The technically optimal solution would be the storage of data in a silicon chip. The most common form of electronic data carrying device in use in everyday life is the chip card based upon a contact field (telephone chip card, bank cards). The mechanical contact used in the chip card is often impractical. A contactless transfer of data between the data carrying device and its reader is far more flexible. In the ideal case, the power required to operate the electronic data carrying device would also be transferred from the reader using contactless technology.

An RFID system is made up of two components:

* The *transponder*, which is located on the object to be identified,
* The detector or *reader*, which, depending upon design and the technology used, may be a read or write/read device.

A reader typically contains a high frequency module (transmitter and receiver), a control unit and a coupling element to the transponder. In addition, many readers are fitted with an additional interface to enable it to forward the data received to another system. The transponder, which represents the actual data carrying device of an RFID system, normally consists of a coupling element and an electronic microchip. When the transponder, which does not usually possess its own voltage supply (battery), is not within the response range of a reader it is totally passive. The transponder is only activated when it is within the response range of a reader. The power required to activate the transponder is supplied to the transponder through the contact less coupling unit as is the timing pulse and data.

RFID FREQUENCIES

RFID tags and readers are required to be tuned into the same frequency to enable communications. RFID systems can use a variety of frequencies to communicate. Radio waves work and act differently at different frequencies, a frequency for a specific RFID system is often dependant on its application. High frequency RFID systems (850 MHz to 950 MHz and 2.4 GHz to 2.5 GHz) offer transmission ranges of more than 90 feet. The wavelengths in the 2.4 GHz range are absorbed by water, which includes the human body, and it has few limitations.

TYPES OF RFID IN THE CONTEXT OF FREQUENCY

1. High Frequency (HF), Long range applications

Long-range RFID applications operate within one meter from the reader and typically involve a large number of tag transponders (100+), requiring reliable anti-collision mechanisms and full interoperability. The products comply with ISO 15693 standard.

Long-range common features:

* 13.56 MHz long-range RFID memory
* ISO 15693 and ISO 18000-3 compliant
* 64-bit unique identifier (UID)
* EAS through AFI
* Anti-collision
* 40 years data retention
* 1 million read/write cycles
* Supports high data rate operations (53 kbit/s)

2. Low Frequency (LF), Short-range Applications

Few vendors offer a wide range of reader ICs and tag ICs for contactless short-range applications such as transport, industrial identification and anti-counterfeiting. These short-range applications use transponders embedded in plastic cards or paper tickets developed through partners. Transaction security is ensured by an optional state-of-the-art anti-clone algorithm. The products comply with the ISO 14443-B standard, 13.56 MHz carrier frequency. Short-range common features are as following:

* 13.56 MHz long-range RFID memory
* ISO 14443-2 type B fully compliant
* ISO 14443-3 type B frame format compliant
* 64-bit unique identifier (UID)
* Anti-collision

* 40 years data retention
* 1 million read/write cycles
* 2 x 32-bit countdown only counters

COMMON USES OF RFID

RFID systems are used in number of applications, from clothing tags to missiles to pet tags to food. It is used where a unique identification system is needed. The tag can carry information such as name, address, instructions etc. Here are a few examples of how RFID technology is being used in everyday places:

* RFID systems are being used in some hospitals to track a patient's location, and to provide real-time tracking of the location of doctors and nurses in the hospital. In addition, the system is used to track the whereabouts of expensive and critical equipment.
* RFID chips for animals are extremely small devices injected via syringe under skin. When scanned, tag can provide information relevant to the animal history and its owner's information.
* RFID in retail stores offer real-time inventory tracking that allows companies to monitor and control inventory supply at all times.

ABOUT RFID TAGS

Most RFID tags contain at least two parts. One is an integrated circuit for storing and processing information, modulating and demodulating a radio-frequency (RF) signal, and other specialized functions. The second is an antenna for receiving and transmitting the signal. Until recently, the high cost of RFID has limited its use. For many applications, such as tracking parts for just-in-time manufacturing, companies could justify the cost of tags—sometimes a dollar or more per tag—by the savings an RFID system could generate. Many times these tags are re-used or have a long life cycle staying attached to the same object for a long time. Cost has been an obstacle for adoption for tracking goods in "open" supply chains (tracking beyond, or outside, a company). Tags built into pallets could be re-used, and some companies are looking to develop ways to recycle tags. There are a variety of groups defining standards and regulating the use of RFID, including: International Organization for Standardization (ISO), International Electrotechnical Commission (IEC), ASTM International, DASH7 Alliance, and EPCglobal. RFID has many applications; for example, it is used in enterprise supply chain management to improve the efficiency of inventory tracking and management.

DEVELOPMENT OF RFID TECHNOLOGY

RFID is an automated data-capture technology that can be used to electronically identify, track, store, and analyze information associated with any physical object using a wireless tag. The roots of RFID can be traced back to World War II. The Germans, Japanese, Americans and British were all using radar to warn of approaching planes while they were still miles away. Radars had been discovered in 1935 by Scottish physicist Sir Robert Alexander Watson-Watt. Mario W. Cardullo received the first U.S. patent for an active RFID tag rewritable memory on January 23, 1973. That same year, Charles Walton, a California entrepreneur, received a patent for a passive transponder used to unlock a door without a key. It was a card with an embedded transponder communicated a signal to a reader near the door. When the reader detected a valid identity number stored within the RFID tag, the reader unlocked the door.

Today RFID systems have the ability to identify and track assets wirelessly. The wireless aspect allows the identity and location of an asset to be determined around corners and through walls. With an appropriately configured RFID system, asset movement and asset location can be determined automatically and instantaneously. Various RFID systems have unique features and constraints involving read range, read reliability, size, power, interference issues, material properties, and site geometries. The concept of RFID can be simplified to that of an electronic barcode. First emerging in the 1980s, RFID was primarily used to track objects in industrial environments where barcodes were unable to sustain the harsh surroundings. Today in 2011, RFID is being used to authenticate official memorabilia, track proprietary assets, automate access control and, since the late 1990s, manage inventory and theft in libraries.

RFID HISTORY

RFID technology has been around since the 1920s. MIT developed the technology to allow robots to "talk" to each other. In 1939, the IFF (identification, friend or foe) transponder was developed. The IFF transponder was used in aircraft to identify themselves as friend or foe to other aircraft. The British used this technology during WWII to identify their aircraft. The RFID History starts with energy as the source of RFID. It would take more than 14 billion years discovered how to harness electromagnetic energy in the radio region, and to apply this knowledge to the development of RFID. From the 1600 to 1800 was an explosion of observational knowledge of electricity, magnetism and optics accompanied by a growing base of mathematically related observations. And, one of the early and well known pioneers of electricity in the 18th Century was Benjamin Franklin.

1. RFID early beginnings

While RFID history can be traced back to the beginnings of radio, and earlier, it really started to come to life when radio technology was used to passively identify objects. The first experiments with radio used in this way were conducted by Sir Robert Watson-Watt when he invented radar. This system used a single transmitter receiver which transmitted a high power signal, and the aircraft (in the first instances) passively reflected a small amount of the radio energy back which was received by the receiver in the radar system. Radar only served to identify the presence of an object, and its size. As mentioned earlier, it was of particular importance to be able to identify an aircraft to see whether it was "Friend or Foe." As a result Watson-Watt went on to develop a system known as IFF - Identification Friend or Foe. A transponder was placed on each plane and when interrogated by a ground station, it responded with a code to identify it. This is the basis of an RFID system.

2. RFID Today

RFID systems are relatively simple. The system usually consists of tags, tag readers, and tag programmers. The tags can be either active or passive. Passive RFID tags do not supply their own power. They are powered by the signal sent from the tag reader. Because these tags do not supply their own power, communication with them needs to be short and usually does not transmit much data (usually just an ID code). The range for transmission is from about 10mm to about 5 meters. Passive tags can be very small (0.4mm x 0.4mm) and not very expensive (roughly $0.40). Active tags supply their own power. They have a much larger range (up to tens of meters) and are larger (about the size of a coin). They are also able to store more information. There are four different kinds of tags in use, categorized by their radio frequency: low frequency (between 125 to 134 KHz), high frequency (13.56 MHz), UHF (868 to 956 MHz), and microwave (2.45 GHz). Low frequency RFID tags are commonly used for pet tracking, automobile key-and-lock, and anti-theft systems. These tags can be implanted into pets so the pet can be returned to their owner if lost or stolen.

High frequency RFID tags are used for applications such as library book tracking, pallet tracking, building access control, airline baggage tracking, and apparel item tracking. These tags are being used to replace magnetic stripe cards in ID badges in places of business. These tags only need to be held within a certain distance to the reader for identification instead of needing to be swiped. UHF RFID tags are used in pallet and container tracking and truck and trailer tracking in shipping yards. Microwave RFID tags are used in long range access to vehicles.

3. How RFID created its niche!

RFID was developed out of the radar experiments and development during the Second World War. The actual date of invention is 1948 but this was followed by decades of development and experimentation before commercial applications were implemented. It's generally said that the roots of radio frequency identification technology can be traced back to World War II. The Germans, Japanese, Americans and British were all using radar—which had been discovered in 1935 by Scottish physicist Sir Robert Alexander Watson-Watt—to warn of approaching planes while they were still miles away. The problem was there was no way to identify which planes belonged to the enemy and which were a country's own pilots returning from a mission. The Germans discovered that if pilots rolled their planes as they returned to base, it would change the radio signal reflected back. This crude method alerted the radar crew on the ground that these were German planes and not Allied aircraft (this is, essentially, the first passive RFID system). Under Watson-Watt, who headed a secret project, the British developed the first active identify friend or foe (IFF) system. They put a transmitter on each British plane. When it received signals from radar stations on the ground, it began broadcasting a signal back that identified the aircraft as friendly. RFID works on this same basic concept. A signal is sent to a transponder, which wakes up and either reflects back a signal (passive system) or broadcasts a signal (active system). It wasn't until the 1970's that RFID implementations started to occur. At this time the cost of the tags was prohibitive and use was limited.

Advances in radar and RF communications systems continued through the 1950s and 1960s. Scientists and academics in the United States, Europe and Japan did research explaining how RF energy could be used to identify objects remotely. Companies began commercializing anti-theft systems that used radio waves to determine whether an item had been paid for or not. Electronic article surveillance tags, which are still used in packaging today, have a 1-bit tag. The bit is either on or off. If someone pays for the item, the bit is turned off, and a person can leave the store. But if the person doesn't pay and tries to walk out of the store, readers at the door detect the tag and sound an alarm.

4. The First RFID Patents

Mario W. Cardullo claims to have received the first U.S. patent for an active RFID tag with rewritable memory on January 23, 1973. That same year, Charles Walton, a California entrepreneur, received a patent for a passive transponder used to unlock a door without a key. Walton licensed the technology to Schlage, a lock maker, and other companies. The U.S. government was also working on RFID systems. In the 1970s, Los Alamos National Laboratory was asked by the Energy Department to develop a system for tracking nuclear materials. A group of scientists came up with the concept of putting a transponder in a truck and readers at the gates of secure facilities. The gate antenna would wake up the transponder in the truck, which would respond with an ID and potentially other data, such as the driver's ID. This system was commercialized in the mid-1980s when the Los Alamos scientists who worked on the project left to form a company to develop automated toll payment systems. These systems have become widely used on roads, bridges and tunnels around the world.

At the request of the Agricultural Department, Los Alamos also developed a passive RFID tag to track cows. The problem was that cows were being given hormones and medicines when they were ill. But it was hard to make sure each cow got the right dosage and wasn't given two doses accidentally. Los Alamos came up with a passive RFID system that used UHF radio waves. The device drew energy from the reader and simply reflected back a modulated signal to the reader using a technique known as backscatter. Later, companies developed a low-frequency (125 kHz) system, featuring smaller transponders. A transponder encapsulated in glass could be injected under the cow's skin. This system is still used in cows around the world today. Low-frequency transponders were also put in cards and used to control the access to buildings.

By the 1980's significant development had occurred and in 1987, Norway implemented the first successful toll collection system. Toll systems were the breakthrough product for RFID. In just a few years, toll systems and government toll collection agencies had spread across the United States fueling the development of the system. We now bring us forward to libraries again. During the 1990's the proliferation of competing systems and radio frequencies employed created the need for standards and interoperability. Most warehousing operations can use large readers which use low radio frequencies. Libraries need the higher frequency waves to allow for smaller, less powerful and portable readers. As complexities and uses increased, standards were developed to allow systems to work together. The final piece of the RFID puzzle was the miniaturization of the tags and the improvements in manufacturing. These two developments dropped the price of the tags to a level that allowed for wide-spread use of the system. The first Library suppliers started to market their systems in the mid 1990's.

5. Technology background

In 1945 Léon Theremin invented an espionage tool for the Soviet Union which retransmitted incident radio waves with audio information. Sound waves vibrated a diaphragm which slightly altered the shape of the resonator, which modulated the reflected radio frequency. Even though this device was a covert listening device, not an identification tag, it is considered to be a predecessor of RFID technology, because it was likewise passive, being energized and activated by electromagnetic waves from an outside source. Similar technology, such as the IFF transponder invented in the United Kingdom in 1915, was routinely used by the allies in World War II. Transponders are still used by most powered aircraft to this day. Another early work exploring RFID is the landmark 1948 paper by Harry Stockman, titled "Communication by Means of Reflected Power" (Proceedings of the IRE, pp 1196–1204, October 1948). Stockman predicted that "... considerable research and development work has to be done before the remaining basic problems in reflected-power communication are solved, and before the field of useful applications is explored."

Mario Cardullo's device in 1973 was the first true ancestor of modern RFID, as it was a passive radio transponder with memory. The initial device was passive, powered by the interrogating signal, and was demonstrated in 1971 to the New York Port Authority and other potential users and consisted of a transponder with 16 bit memory for use as a toll device. The basic Cardullo patent covers the use of RF, sound and light as transmission media. The original business plan presented to investors in 1969 showed uses in transportation (automotive vehicle identification, automatic toll system, electronic license plate, electronic manifest, vehicle routing, vehicle performance monitoring), banking (electronic check book, electronic credit card), security (personnel identification, automatic gates, surveillance) and medical (identification, patient history).

An early demonstration of *reflected power* (modulated backscatter) RFID tags, both passive and semi-passive, was performed by Steven Depp, Alfred Koelle, and Robert Freyman at the Los Alamos National Laboratory in 1973. The portable system operated at 915 MHz and used 12-bit tags. This technique is used by the majority of today's UHFID and microwave RFID tags. The first patent to be associated with the abbreviation RFID was granted to Charles Walton in 1983.

The largest deployment of active RFID is the US Department of Defense use of Savi active tags on every one of its more than a million shipping containers that travel outside of the continental United States (CONUS). The largest passive RFID deployment is the Defense Logistics Agency (DLA) deployment across 72 facilities implemented by ODIN who also performed the global roll-out for Airbus consisting of 13 projects across the globe. Over time, companies commercialized 125 kHz systems and then moved up the radio spectrum to high frequency (13.56 MHz), which was unregulated and unused in most parts of the world. High frequency offered greater range and faster data transfer rates. Companies, particularly those in Europe, began using it to track reusable containers and other assets. Today, 13.56 MHz RFID systems are used for access control, payment systems (Mobile Speedpass) and contactless smart cards. They're also used as an anti-theft device in cars.

There was some activity in the 1960's and 1970's in the application of RFID technology, with a fundamental patent filed in 1973 by Charles Walton, a former IBM researcher who left IBM to form his own company, Proximity Devices, in Sunnyvale, California. It is the fundamental idea behind the access cards used today.

In the early 1990s, IBM engineers developed and patented an ultra-high frequency (UHF) RFID system. UHF offered longer read range (up to 20 feet under good conditions) and faster data transfer. IBM did some early pilots with Wal-Mart, but never commercialized this technology. When it ran into financial trouble in the mid-1990s, IBM sold its patents to Intermec, a bar code systems provider. Intermec RFID systems have been installed in numerous different applications, from warehouse tracking to farming. But the technology was expensive at the time due to the low volume of sales and the lack of open, international standards.

UHF RFID got a boost in 1999, when the Uniform Code Council, EAN International, Procter & Gamble and Gillette put up funding to establish the Auto-ID Center at the Massachusetts Institute of Technology. Two professors there, David Brock and Sanjay Sarma, had been doing some research into the possibility of putting low-cost RFID tags on all products made to track them through the supply chain. Their idea was to put only a serial number on the tag to keep the price down (a simple microchip that stored very little information would be less expensive to produce than a more complex chip with more memory). Data associated with the serial number on the tag would be stored in a database that would be accessible over the Internet. Sarma, then an engineering professor at the Massachusetts Institute of Technology (MIT), along with a colleague, David Brock, were working on a robotics problem - how to have a machine recognize and respond to items in its environment. What if, instead of having a robot attempt to "recognize" objects optically, what if objects, with the use of an electronic marker, could identify themselves to a robot? Using a simplified marker, or tag, a robot could be informed of the presence of an item and then use its identifier to access a database to obtain specific information on the item for reference.

In 1999, Sarma, Brock, and Kevin Ashton co-founded

the Auto-ID Center at MIT, to explore methods by which RFID technology could be used in commercial applications. The Auto-ID Center helped to develop the concept of using RFID as a networked technology during the time when many saw the power of using the internet as a foundation for applications. Research at the Auto-ID Center created a "new" interest in RFID with considerable resources made by industry to capitalize on Stockman's and Sarma's ideas. They essentially changed the way people thought about RFID in the supply chain. Previously, tags were a mobile database that carried information about the product or container they were on with them as they traveled. RFID was turned into a networking technology by linking objects to the Internet through the tag. For businesses, this was an important change, because now a manufacturer could automatically let a business partner know when a shipment was leaving the dock at a manufacturing facility or warehouse, and a retailer could automatically let the manufacturer know when the goods arrived.

Between 1999 and 2003, the Auto-ID Center gained the support of more than 100 large end-user companies, plus the U.S. Department of Defense and many key RFID vendors. It opened research labs in Australia, the United Kingdom, Switzerland, Japan and China. It developed two air interface protocols (Class 1 and Class 0), the Electronic Product Code (EPC) numbering scheme, and a network architecture for looking up data associated on an RFID tag on the Internet. The technology was licensed to the Uniform Code Council in 2003, and the Uniform Code Council created EPCglobal, as a joint venture with EAN International, to commercialize EPC technology. The Auto-ID Center closed its doors in October 2003, and its research responsibilities were passed on to Auto-ID Labs. Some of the biggest retailers in the world—Albertsons, Metro, Target, Tesco, Wal-Mart—and the U.S. Department of Defense have said they plan to use EPC technology to track goods in their supply chain. The pharmaceutical, tire, defense and other industries are also moving to adopt the technology. EPCglobal ratified a second-generation standard in December 2004, paving the way for broad adoption.

6. RFID is not a "new" technology

It is fundamentally based on the study of electromagnetic waves and radio, which was rooted in the 19th century work of Michael Faraday, Guglielmo Marconi and James Clerk Maxwell. The concept of using radio frequencies to reflect waves from objects dates back as far as 1886 to experiments conducted by Frederick Hertz. Radar as we know it was invented in 1922 and its practical applications date back to World War II.

Technology advances in a variety of different fields—computers, radio, radar, supply chain management, transportation, quality management, and engineering—have made RFID technology more useful with applications in asset management, payments, ticketing, livestock tracking and transportation. The US government helped advance the RFID technology during this time, as the Los Alamos National Laboratory in New Mexico performed research and development in radio frequency technology with research in livestock tracking applications, railroad applications and toll road applications. Exxon/Mobil introduced the Speedpass transponder device in 1997 to allow drivers to make credit gasoline purchases by waving a key fob with a tiny transponder in front of the gas pump. The program has an estimated 7 million subscribers.

SIGNIFICANCE

1. Automated materials handling

RFID technology enables automated materials handling. This includes conveyor and sorting systems that can move library materials and sort them by category into

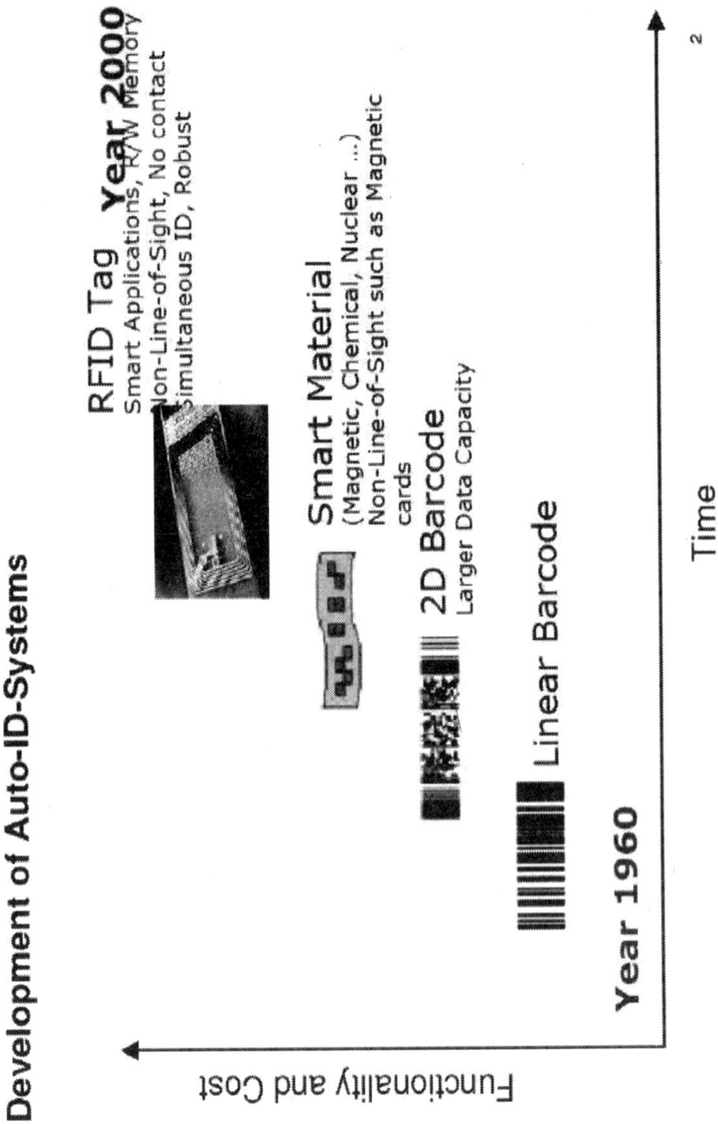

Development till 2000, Source: RFID - Technology and Applications in Libraries. Nils Rinaldi. Ecole Polytechnique Fédérale de Lausanne Station 5, 1015 Lausanne (Switzerland), IFLA-PAC Conference, 26th March 2008.

separate bins or onto separate carts. This significantly reduces the amount of staff time required to ready materials for re-shelving. Given the high cost of this kind of equipment, this application has not been as widely used. The history of RFID has shown a steady development in RFID technology. Having its routes in the earliest days of electrical science and then radio, RFID has come out of developments such as radar and IFF. Now RFID is a technology in its own right which is widely used and showing massive benefits to industry and society as a whole.

2. High-speed inventory management

A unique advantage of RFID systems is the ability to scan items on the shelves without tipping them outward from the shelves or removing them. Library personnel can use a hand-held RFID reader, moving it rapidly across a shelf of books, to read all of the unique identification information for the items on a shelf. Using wireless technology, it is possible not only to update the inventory, but also to identify items which are out of proper order. This helps to identify items that have been mis-placed in the stacks or shelves.

A HISTORY OF CONSUMER BENEFITS FROM RFID

Application of RFID technology in the society is more than 60 years old. A primitive form was first used in World War II to distinguish friendly from enemy aircraft. Later shops began using the technology in anti-theft systems installed at doors. Then came "keyless" access systems based on RFID. These were first developed in the 1970s and 1980s. Research into low-cost tagging first made RFID an option for widespread use in the late 1990s. At that point companies began to experiment with applications which allow user to pay for bus ticket with his/her mobile phone or which identify counterfeit and therefore potentially dangerous medicinal products. About the same time, the Electronic Product Code (EPC) made its appearance. This consists of a lengthy number similar to the barcode and was created by a consortium of corporations and university labs in anticipation of the global use of RFID. The EPC Network was created by EPCglobal Inc. (a subsidiary of GS1).

DISADVANTAGES OF RFID SYSTEMS

Tagging every consumer product would bring untold logistical benefits, but in a world where privacy, or rather the lack of it, is a growing concern, this master plan of world domination is going to encounter some resistance. How does one even attempt to bring about this type of global change? What are the DOs and DON'Ts, and can therefore provide clues to other smaller, organizational changes?

RFIDs fire the imagination; they encourage speculation, and feed both reasonable and unreasonable privacy fears. RFID announcements must be made with this context in mind, they must address these concerns head on from the start, to do otherwise allow these speculations to grow and become as contagious, and just as impervious to reason, as urban legends. To date, there is a dearth of RFID industry response to the privacy issues.

There are some uses of RFID which will fail because of consumer backlash. The fault of the failure will not lie with the consumer, but with the company who insisted, against the court of public opinion, on using RFID in that manner. Bar codes are usually scanned at the store, but not after purchase. But RFID transponders are, in many cases, forever part of the product, and designed to respond when they receive a signal. Imagine everything is "numbered, identified, catalogued, and tracked." RFID raises legitimate concerns. RFID is not totally rejected by the consumer. The open flow of information during a Change project is a key to a successful implementation.

Here's a key Change lesson. To overcome this type of resistance is to find a way with information, involvement and innovation. The privacy issue was obvious from the beginning. Few other aspects are as following:

1. **High cost:** The major disadvantage of RFID technology is its cost. The readers and sensors used to read the RFID tags are comparable in cost to the components of a typical EM or RF theft detection system. The RFID tags cost more than barcodes and EAS tags.

2. **Vulnerability to compromise:** It is possible to compromise an RFID system by wrapping the protected material in ordinary household aluminum foil to block the radio signal. Bringing household foil into a library using RFID would represent premeditated theft, just as bringing a magnet into a library using EM technology would be. RFID tags may be removed by patrons. EM strips are concealed in the spines or the gutters of books and are more difficult to find and remove; while RFID tags are typically affixed to the inside back cover and are exposed for removal. Libraries may imprint the RFID tags with its logo and make them appear to be bookplates, or it can put a printed cover label over each tag.

3. **Exit sensor problems:** While the short-range readers used for circulation charge and discharge and inventorying appear to read the tags 100 percent of the time, the performance of the exit sensors is more problematic. They must read tags at up to twice the distance of the other readers.

4. **Invasion of User Privacy:** There may be a perception among some that RFID is a threat to patron privacy. That perception often revolves around the idea that the RFID tags contain patron information or that the RFID tags might be read by a device other than devices authorized by the user organization. The vast majority of the tags installed in materials contain only the item ID, usually the same number that previously has been stored

on a barcode. The link between borrower and the borrowed material is maintained in the circulation module of the automated system. When we talk in the context of library, the tag consists of information about the item, including holding location, call number, and rarely author/ title. It is important to educate organizational staff and users about the RFID technology. This may include notices regarding the use of RFID technology, information about how it uses RFID technology and a description of the data stored on RFID tags. An educated user is the best defense against perceived privacy issues.

RFID, its application, standardization, and innovation are constantly changing. Its adoption is still relatively new. There may be many features of the technology that are not well understood by the general populace. Developments in RFID technology continue to yield larger memory capacities, wider reading ranges, and faster processing. It's highly unlikely that the technology will ultimately replace bar code - even with the inevitable reduction in raw materials coupled with economies of scale, the integrated circuit in an RF tag will never be as cost-effective as a bar code label. However, RFID will continue to grow in its established niches where bar code or other optical technologies aren't effective. If some standards commonality is achieved, whereby RFID equipment from different manufacturers can be used interchangeably, the market will very likely grow exponentially. Market expectations that widespread deployment of RFID for item-level tagging would by now be commonplace are being viewed as exaggerated by some industry watchers.

Could it be that the early disappointments with expectations for item-level tagging now mark a turning point for RFID? RFIDsec and the global partners believe that RFID technology is now more important than ever. New possibilities are based on the new generation of "RFID Solutions" or "RFID 2.0." Although the term "Web 2.0" suggests a new version of the Web, to change in the ways that developers and end-users can now use the Web. It provides opportunity, moving from pure content provision to collaborative development of content. With RFID, it is different. In addition to changes in the ways that developers and end-users can use RFID technology, "RFID 2.0" will require enhanced security in the RFID Tags and in the communication between Reader and RFID Tag.

2

An Overview of RFID System Elements

INTRODUCING RFID

RFID is defined as a method of identifying unique items using radio waves (http://www.rfidjournal.com). RFID technology allows the automatic collection of product, place, and time and transaction data quickly and easily without human intervention. An RFID system includes a reader, a transponder, and their associated antennas. The reader transmits the radio signal, through its antenna, that the transponder receives via its own antenna. The transponder converses with the reader to verify and exchange the data. Once the reader receives and eventually verifies the data, it sends it to a computer for processing and management. RFID readers are automatic locks, fixed or mobile hand held scanners. They are usually connected to a computer and serve the same purpose as a barcode scanner. The transponder, also called tag, smart card or smart label, consists of a chip containing a processor and a receiver, and an antenna to broadcast and receive data via radio frequency. Dependent on the installation size of the antenna and the air interface protocol, the coverage reaches up to several meters. In contrast with the barcode, the transponder can be read without direct visibility and is contact-free. Additionally, transponders can store more information and are safer in terms of staining or abrasion. A huge advantage of RFID is the parallel data collection: a RFID reader can read up to 200 tags. The antenna connected to the RFID reader activated the RFID tag and transfers data by emitting wireless pulses.

A typical RFID system is made up of several system elements:

1) Tags
2) Provisioning Devices (Printing/Encoding Devices)
3) Interrogating Devices (Readers, Reader/Writers)
4) Antennas
5) RF Signal (Using specific frequencies and air interface protocols)
6) Backend Processing Systems

These are discussed and explained as following:

RFID TAGS

RFID tags and RFID tagging techniques are essential to the whole operation of the Radio Frequency Identification system. In general the RFID tags or RFID transponders are small low cost items that can be attached to items that need tracking or other forms of data collection to be performed. The RFID tags are the small plastic elements. However tags take a variety of forms and RFID tagging can be accomplished in a variety of ways. Radio Frequency Identification tags are made to be as simple as possible and they contain comparatively little in the way of electronics. Fundamentally they comprise two main elements:

Electronics circuitry: The electronics within an RFID tag are kept to the minimum so that costs are minimized and power levels are kept as low as possible.

Antenna: The antenna takes the largest amount of space. It must be able to operate satisfactorily at the frequency of operation. With wavelengths being smaller at higher frequencies (especially UHF and microwave), this makes antennas for these frequencies much more efficient.

1. RFID tag types

RFID tags or RFID transponders can take a variety of forms. There are three main categories into which they fall:

Passive: Passive RFID tags are by far the most common. They do not contain any power and receive this from the RFID reader. This is sufficient to power any device in the RFID tag and reply with the required data. The so called RFID smart tags, or RFID smart labels are all passive.

Semi-passive : This form of RFID tag uses a battery to supply the internal operation of the tag, but relies on the RFID reader to supply the power to transmit the signal to the reader.

Active : An active RFID tag is one in which battery power is used to supply power to the electronics. This enables greater distances to be achieved as the tag is not dependent upon the received power to provide a reflected signal, and the control and processing circuits can be more sophisticated as in the case of the semi-passive RFID tag.

2. Another Classification of RFID Tags: Read only and read-write RFID tags

RFID tags may be able either to perform as a read only RFID tag, or they may be a read-write Radio Frequency Identification tag. In view of the cost of manufacturing different types against the quantities made and the differences between the two, most RFID tags today are the read-write variety. For applications where only a read function is required, the write ability is not used. Read only radio frequency identification tags are typically programmed in the factory. Data included will be a unique identifier and other specified data that cannot be changed. Read-write RFID tags normally contain an area where data cannot be altered - this is often a segregated secure read-only area in the memory. Again this will include a unique identifier, and other data that may be required. The writeable area can then be used to contain data that may be required. For example if the RFID tag is used with a container, it can contain details of the container contents, etc. This area of memory within the RFID tag can be re-written many times.

3. RFID tag storage and processing

One important area and function of the RFID tag is the area that handles the information storage and processing. RFID tags range vastly in their capabilities as some do not have their own power. They rely on the received signals to provide any power and this limits their abilities. Other RFID tags with their own battery power are able to carry out far more sophisticated tasks.

There are several types of RFID tag that may be used:

One-bit EAS RFID tags : EAS (Electronic Article Surveillance) tags are commonly found in shops and stores to prevent theft. EAS tags are often termed "1 bit" tags. The reason for this is simply that they are only designed to communicate one bit of information, i.e. their presence. They are widely used in anti-theft measures in shops and stores. If the RFID tag is present and active, then it means that the item has not been through the checkout. If they have been passed through the checkout the RFID tag is either deactivated or removed.

Because of their use, EAS tags are used in their millions and possibly the most widely used form of RFID tag. They do not have any memory or other chips as these would make them too expensive. Coupling used for these tags is generally inductive or backscatter. The tags simply consist of a resonant circuit, and the reader is able to detect their presence.

RFID smart labels : Smart labels are simple RFID tags that are embedded in an adhesive paper label. The advantage of this form of tag is that they can be used by RFID and barcode readers as well as having the option for human readable characters. They can be used in areas where the end product may enter one of a number of scenarios where the form of reader is not known. For example, retail outlets a product may be shipped to may have either a barcode reader or an RFID reader, and outlets will have different options. Therefore to cover all eventualities a combined RFID and barcode tag is printed.

SAW RFID tag : SAW (Surface Acoustic Wave) tags form a half way house between the very basic 1-bit RFID tags. Even few more advanced tags are available. The SAW RFID tags operate in the microwave region using backscatter techniques. They do not have a processor; they can be encoded at point of manufacture with a number. This number is limited by the technology but may be up to 32 or 64 bits.

Smart card tags : Smart card tags are different to smart labels. Advanced smart card tags are used for many applications, and in particular where secure communications is required. For example, for transactions involving finance. These cards may have complicated processors on board along with sufficient memory. When using these cards, there is a balance to be made between functionality and cost. It requires to be taken at the outset of the design and needs to be carefully balanced.

Although RFID tags may appear to be more straightforward or simple element within an RFID system, this may not be the case as considerable ingenuity. Careful design is required to ensure that the RFID tags perform correctly while being capable of being manufactured to a very low cost. It should be within constraints of size, weight, form factor and also reliability. While most RFID tags are very cheap to manufacture, this hides the design behind them. RFID tags generally consist of a microchip, a strap, an inlay, an antenna and packaging. Most RFID tags contain a chip, but some tags are chipless. RFID tags have a microchip attached to an antenna that is packaged in a way so that the tag may be applied to an object. A RFID tag receives signals from a reader and sends signals to a reader. The tag contains a unique serial number, but may have more information. RFID tags come in many forms and may be applied to an object in many ways.

RFID tags may operate under a wide variety of environments (harsh environmental conditions, a wide range of temperatures or in the presence of materials that distort or attenuate radio waves). Future research regarding RFID tags involves power conservation, tag power sharing, tag-to-tag communication and other research to optimize the performance and cost of tags for applications in which they operate. An RFID tag is the most important link in any RFID system. It has the ability to store information relating to the specific item to which they are attached and rewrite again without any requirement for contact or line of sight. Data within a tag usually may provide identification for an item, proof of ownership, original storage location, loan status and history.

RFID Tag consists of an integrated circuit and an antenna combined to form a transponder. RFID tags collect the energy to operate from a Radio Frequency (RF) field emitted by a reader device; therefore they do not need a battery. When energized by a radio signal from a fixed position reader or handheld scanner, the tag returns the stored information in order that the item to which it is attached can be easily located. The chip also has a "multi-read" function, which means that several tags can be read at once. RFID tags have been specifically designed to be affixed into library media, including books, CDs,

DVDs and tapes. It is thin, flexible and thus can be laminated between paper and plastic. With special method to attach to books, patron is totally unaware that the tag is there.

4. Key Benefits

- No line of sight needed
- Allows to check-out and check-in several items simultaneously
- Information directly attached to product
- Performing both identification and antitheft in one single operation
- Different shape and sizes available
- Able to tag almost anything
- Accelerate scanning and identifying

5. How Tags Communicate

The communication process between the reader and the tag is by wireless. The major differences between the different types of waves are the distances covered by one cycle of the wave and the number of waves that pass a certain point during a set time period. It is necessary to understand that the wavelength is the distance covered by one cycle of a wave and the frequency is the number of waves passing a given point in one second. For any electromagnetic wave, the wavelength multiplied by the frequency equals the speed of light. The frequency of an RF signal is usually expressed in units called hertz (Hz). One Hz equals one wave per second. Basically what happens is that when the reader is switched on it starts emitting a signal at the selected frequency band, for example, in library HF is used with 13.56 MHz. Any corresponding tag in the vicinity of the reader will detect the signal and use the energy from it, to wake up and supply operating power to its internal circuits. Once the tag has decoded the signal as valid, it replies to the reader and indicates its presence by modulating (affecting) the reader field.

6. Anti-collision

If many tags are present (such as in a row of books) then they will reply at the same time, which at the reader end is seen as a signal collision and an indication of multiple tags. The reader manages this problem by using an anti-collision algorithm designed to allow tags to be sorted and individually selected. The number of tags that can be identified depends on the frequency and protocol used, and typically ranges from 50 tags for HF and up to 200 tags for UHF. Once a tag is selected the reader is able to perform a number of operations such as read the tags identifier number, or in the case of a read/write tag write information to it. After finishing dialoging with the tag the reader can then either remove it from the list, or put it on the stand by until a later time. This process continues under the control of anti-collision algorithm until all tags have been selected. In fact very real challenges for the ICs' exist such as achieving very low power consumption, managing noisy RF signals and keeping within strict emission regulations. Other important function of the circuit is to allow the chip to transfer power from the reader signal field, and convert it via a rectifier into supply voltage. The chip clock is also normally extracted from the reader signal. Most RFID tags contain a certain amount of NVM (non-volatile memory) like EEPROM in order to store data.

The amount of data stored depends on the chip specification, and can range from just simple identifier numbers of around 96 bits to more information about the product with up to 32Kbits. In 1999 the AUTO-ID centre (now EPC Global) based at the MIT-USA, together with the number leading companies developed the idea of an unique electronic identifier code called the EPC (Electronic Product Code). The EPC is similar in concept the UPC used in barcodes today. Having just a simple code of up to 256 bits would lead to smaller chip size, and hence lower tag cost, which is recognized as the key factor for wide spread adoption of RFID. Like a barcode, the EPC is a 96 bit unique number which is divided into numbers that identify the manufacturer, product, version and serial number.

7. Tag IC's

A single-chip design led to the RFID tag, a small device composed of a chip, an antenna, and an optional power source, that carries a unique identifier. The 1990s witnessed the use of such tags for card-keys, fuel-station payment systems, and automated toll payment. Such tags were typically specialized for a certain class of applications and cost a few dollars each. The tags typically stored application-specific data and were capable of modest processing on-tag.

8. Tag Classes

One of the main ways of categorizing RFID tags is by the capability to read and write data. This leads to the following four classes and EPC global has also defined five classes which are similar to the one below:

Class 0: These are Read only - factory programmed tags. These are simplest type of tags, where the data, which is usually a simple id number is written only once into the tags during manufacture. The memory is then disabled from any further updates. Class 0 is also used to define a category of tags called EAS or anti-theft devices which have no id, and only announce their presence when passing through an antenna field.

Class 1: These are Write Once Read Many (WORM) Factory or user programmed tags. In this case tag is manufactured with no data written in to the memory. Data can then either be written by manufacturer or by the user - one time. Following this no further writes are allowed and the tag can only be read. Tags of this type usually act as simple identifiers.

Class 2: These are Read-Write tags. This most flexible type of tag, where users have access to read and write data into the tags memory. They typically used as data loggers, and therefore contain more memory space than what is needed for just a simple id number.

Class 3: These are Read-Write (with on board sensors) tags. These tags contain on board sensors for recording parameters like temperature, pressure and motion, which can be recorded by writing into the tags memory. As sensor readings must be taken in the absence of a reader, the tags are either semi-passive or active.

Class 4: These are Read-Write (with Integrated Transmitters) tags. These are like miniature radio devices which can communicate with other tags and devices without the presence of the reader. This means that they are completely active with their own battery power source.

9. RFID Tag Design

RFID vendors supply standard production tags/ transponders for all RFID frequencies. The RFID vendors also provide full customization services for the specialized tag requirements. All RFID inlays do not perform the same even though they appear to look identical when viewed with the naked eye. Some vendors can supply with standard production inlays encapsulated or packaged to the needs, and the RFID vendors can also supply organization/user with special higher performance inlays and finished tags. While many people are familiar with some of the inlay manufacturers, most people are not aware that an inlay can be produced with a copper, silver or aluminum antenna. Each of these 3 different antenna materials produces different read results! In addition to the antenna, how the inlay is attached to the antenna and what materials are used to attach the antenna also adds to the readability percentage and read distance equation.

RFID Tags can be tuned to the material or object that the tag is going to be attached to. When organization/

user know that a tag must be attached to a specific surface the RFID vendors can tune the tag order so that organization/user obtains the maximum read range when the tag is attached to the material(s) that organization/ user specify with either the standard manufacturers inlays or the higher performance inlays. If organizational application needs are such that the inlay and encapsulation/packaging requirements of a standard tag will result in too high of a read range, the RFID vendors can de-tune the tag so that organization/user obtain only the read distance that are required. This is especially good when using high read distance inlays and organization/ user need to keep the read range to 1 - 2 Meters (3 - 6 feet) as the standard range of 5 - 7 Meters (16 - 22 feet) would cause too many false reads due to the proximity of other tags.

In addition to the tuning/detuning of standard and high performance inlays certain vendors can also supply full engineering design and prototyping services for custom tags. Vendors can provide custom or boutique tags in 125 kHz., 134.2 kHz., 13.56 MHz. the 860 ~ 960 MHz. range and the 2.4 GHz frequencies. Through the strategic relationships the RFID vendors have access to some of the finest equipment and brightest engineers in the world. Through them the RFID vendors can provide services to engineer and prototype a custom tag consisting of tailored integrated circuits (IC's) and customized antenna coupled with encapsulation/packing materials that meet or exceed the unique RFID requirements.

10. RFID Access Tag

There are wide variety of RFID access tags that organization/user can use to access parking lots, buildings, offices, cabinets and more. A few years ago most RFID access tags were either Low Frequency 125 kHz. or 134.2 kHz. for almost contact reads and HF 13.56 MHz.

for the higher security and longer distance reads (approximately 1 meter). While it is possible to get a foot or so read range from the LF tags the size required in order to do so makes it inconvenient to carry the tag in the pocket or purse.

Since LF is still very popular for access control, organization/user can get a very wide variety of key ring tags (keyfobs) in ABS plastic, PVC or even in leather-Vendors can also supply an ABS keyfob in LF, HF or UHF! Most of the key fob type HF RFID Tags and UHF RFID Tags are used for RFID Door Lock access control for buildings and rooms. When it was first started RFID parking lot access control in the early 90's it had to dig up asphalt and concrete driveways and put in underground antenna that were almost 4 feet square. Today we don't need to do that as vendors can place the antenna in a secured place above ground and the reader can be hidden inside the building. Many companies now want a specific access frequency such as the HF/UHF range for parking lot access where organization/user needs a longer read distance, a Low Frequency (125 kHz. or 134.2 kHz.) short read range for the main entrance and exit doors and another HF frequency vicinity or proximity read range (13.56 MHz.) for higher secured areas.

A few years ago this meant that each person needed a LF RFID tag, and HF RFID tag and then a UHF RFID tag (3 types of tags) for access control and the company used 3 separate systems to control all of that access. Today there are the UHF or 860 ~ 960 MHz. frequency that can be used for the parking lot access with a range of about 7 meters or 22 feet. In addition to that the RFID vendors are now able to have 2 or 3 different RFID chips and frequencies in the same card RFID tag. This is perfect for organizations that have the need for 2 or more different RFID frequencies and want a single access tag. This ISO triple (3) frequency credit card sized RFID tag is made of PVC and is super slim even with all 3 frequencies inside the single card.

Organization/user can print logos or graphics on it, add a magnetic strip to it and even a signature stripe. Looks like a standard credit card but in reality is a highly secured LF, HF and UHF Gen 2 EPC Global card. In many countries today these card RFID tags are being used to pay for food, transportation, groceries, shopping purchases and many other uses, as well as being the main access for the parking lot. RFID vendors can provide organization/user with almost any LF, HF and Gen 2 UHF chip combination that organization/user need for the multiple access control demands. If organization/user really only wants a single frequency ISO card, vendors can also supply. When organization/user wants to have the LF, HF UHF or combination RFID card visible for photo identification and ease of use when organization/user have multiple levels of security and many employees, contractors or visitors coming in and out of secured areas, organization/user can use the tag. This card can have all or any of the ISO card frequencies plus organization/user can have a lanyard type clip so organization/user can have the card attached to a lanyard or an expandable belt or pocket clip. The RFID vendors also have a wide variety of fixed readers in Low Frequency, High Frequency and Ultra High Frequency to support the new or updated access control solution for security.

11. Active and Passive tags:

First basic choice when considering a tag is either passive or semi-passive or active. Passive tags can be read at a distance of up to 4 - 5 m using UHF frequency band, whilst the other types of tags (semi-passive and active) can achieve much greater distance of up to 100m for semi-passive, and several KM for active. This large difference in communication performance can be explained by the following;

* Passive tags use the reader field as a source of energy for the chip and for the communication from

and to the reader. The available power from the reader field, not only reduced very rapidly with distance but is also controlled by the strict regulations, resulting in a limited communication distance of 4 -5 m when using UHF frequency band (860 MHz - 930 MHz) .

* Semi-passive (battery assisted back scatter) tags have built in batteries and therefore do not require energy from the reader field to power the chip. This allows them to function with much lower signal power levels, resulting in greater distance of up to 100meters. Distance is limited mainly due to the fact that tag does not have an integrated transmitter, and is still obliged to use the reader field to communicate back to the reader.
* Active tags are battery powered devices that have an active transmitter onboard. Unlike passive tags, active tags generate RF energy and apply to the antenna. This autonomy from the reader means that they can communicate at the distance of over several KMs.

Table 2.1: Different Tag Classes

class	Known as	Memory	Power source	Applications
0	EAS/EPC	None/EPC-1 bit on/off	Passive	Anti-theft/ID
1	EPC	Read only	Any	Identification
2	EPC	Read-Write	Any	Data logging
3	Sensor tags	Read-write	Semi passive/ active	sensors
4	Smart Dust	Read-write	Active	Ad hoc networking

Source: A Basic Introduction To RFID Technology And Its Use In The Supply Chain. Available At Http://www.Idii.Com/Wp/Laranrfid.Pdf

Defining Active Tag (Active RFID Tag): An RFID tag is an active tag when it is equipped with a battery that can be used as a partial or complete source of power for the tag's circuitry and antenna. Some active tags contain replaceable batteries for years of use; others are sealed units. It is also possible to connect the tag to an external power source). Active RFID tags may have all or some of the following features:

* It is capable to perform independent monitoring and control
* It is capable of initiating communications
* It has the highest data bandwidth
* It has longest communication range of any tag
* It is capable of performing diagnostics
* It may even be equipped with autonomous networking as the tags autonomously determine the best communication path.

The major advantages of an active RFID tag are:

* It can be read at distances of one hundred feet or more, greatly improving the utility of the device
* It may have other sensors that can use electricity for power.

The problems and disadvantages of an active RFID tag are:

* The tag cannot function without battery power, which limits the lifetime of the tag.
* The tag is physically larger, which may limit applications.
* The tag is typically more expensive, often costing $20 or more each
* Battery outages in an active tag can result in expensive misreads.
* The long-term maintenance costs for an active RFID tag can be greater than those of a passive tag if the batteries are replaced.

12. Advantages and disadvantages of active and passive RFID tags: a comparison

When choosing an active or passive tag, it is necessary to determine the requirements and whether an active RFID tag or a passive RFID tag is needed.

Some of the advantages and disadvantages of active and passive RFID tags are given below:

1. Active RFID tags

Advantages

* The range of active RFID tags can be many metres - often 30 metres or more.
* As battery power is required, this may allow much greater functionality to be incorporated.

Disadvantages

* An active RFID tag cannot function without battery power. This limits its lifetime, or requires maintenance.
* RFID active tags are much more expensive to manufacture.
* An active RFID tags will be physically larger than a passive one and this may limit its use. · Battery end of life can result in loss of functionality while the battery is down. This can be seen as a failure within the system

2. Passive RFID tags

Advantages

* A RFID passive tag by definition does not have a battery and their life is almost indefinite.
* Passive RFID tags are simpler to manufacture than active RFID tags and therefore cost considerably less.
* As passive RFID tags are generally simpler, and they do not contain a battery they can be made much

smaller than active ones.

Disadvantages

* The range of passive RFID tags is limited - typically a metre or so.
* As power in RFID passive tags is very limited, sensors are rarely fitted as these will require additional battery power.
* The passive tag remains operable for possibly years after it was initially used - this can have personal privacy implications unless it is deactivated.

Most active RFID tags are used for **location** and **tracking systems**. The active tag transmits its identity on a predetermined basis and with the correct antenna placement organization/user can determine the location of the active tag. At the same time, the **Temperature sensing tags** can be used to store temperature information on a timed basis and send the collected information to the active tag reader. In addition the temperature sensing tag can be set up to set off an alarm condition. The **Motion sensing active RFID tags** can be set up to transmit a signal when moved or signal until movement stops. This is ideal for high value asset control where organization/user want to know if an asset is moving so that organization/user can determine if the movement is justified or an intrusion. It has been seen that active RFID tags can also be used for **patient tracking** in health care facilities. Using a wrist band tag patients movements within a certain location are known and if the patients goes outside of the area organization/user can have an alarm sound or have a health care giver return the patient to the controlled area.

13. RFID Asset Tag

Some vendors provide a wide variety of Asset Tags (LF, HF, UHF and SHF) with different substrates, coatings and backings depending on how the tag is going to be used. A tag designed for a wooden object will not read well if attached to metal and likewise a tag designed for mounting on metal will not read as well as it could if it's attached to wood or plastic. They can provide cost effective asset tags in a variety of materials including polypropylene, polyester, polycarbonate, polyolefin and more. These tags can have logos and other printing on the viewing surface and be mounted using a single side adhesive, foam adhesive or folded hang tag. Vendors have a wide variety of easy to use RFID Asset Tags.

14. RFID Animal Tag

Before organizations/users start tagging animals, livestock and other living creatures such as birds, rabbits, turtles, shell fish, iguanas, monkeys and zebras, they should understand the basics of the ISO RFID standards and how it may or may not affect what they can or can't do legally. Current ISO RFID standards for animal tracking are ISO 11784 and ISO 11785. The air interface standard ISO 14223-1 is an extension of the two standards and is supposed to be implemented in the reader for full ISO RFID compliance.

ISO 11784 describes the structure and data content. For animal use, the content includes the tag manufacturers code, country code, and unique identification number. The country code is based on another ISO standard and the manufacturer's number is assigned by ICAR (International Committee of Animal Recording); the ISO 11785 describes the technical aspects of the communication method between the tag and the reader. The standard can be Half Duplex (HDXS), one way communication or Full Duplex (FDX-A, FDX-B) two way communications. Where:

* FDX-A is a 10 character alpha-numeric 125 kHz. Animal tag
* FDX-B is a 15 character alpha-numeric 134.2 kHz. Animal tag.

Currently the FDX-B standard is patented and in order to use a 134.2 kHz. FDX-B animal tag in North America or any other country that has a patent relationship with United States organization/user must pay royalties as part of the cost. This also applies to 125 kHz. and 134.2 kHz. animal glass tags.

While the FDX-B standard has been adopted by United States and several other countries for Livestock (RFID Cattle)- most countries have not fully adopted the standard and in fact the 125 kHz FDX-A standard is in use in many countries with excellent results and no royalty payments. This difference in standards has resulted in many readers being compliant to 11784/85 FDX-B, 134.2 kHz only and other readers being FDX-A and 125 kHz. There are very few (expensive) readers available that use both 125 kHz. and 134.2 kHz frequencies and HDX, plus FDX-A and FDX-B in accordance to the full ISO standards.

There are no standards in the ISO for Animal Tracking so there can be duplications of the identification number. Counterfeiting can occur using a WMRM (Write Many Read Many) tag. Some vendors offer 125 kHz. RFID Animal Tags and 134.2 kHz. Animal Tags but is not able to sell the 134.2 kHz. glass tags where patent issues may exist. 125 kHz. RFID Animal tags are available at a far more reasonable price than the 134.2 kHz tags. And if organization/user needs to tag creatures, the RFID vendors have excellent 125 kHz. animal Tag solutions that work very well.

15. RFID CD and DVD Tags

CD and DVD control and management has never been easier to implement and obtain excellent results. Some RFID vendors provide HF RFID CD and DVD label tags and long read range UHF RFID CD and DVD label tags. These CD and DVD tags are easy to peel and stick onto virtually any CD or DVD and protect the assets. For example, libraries currently using a HF RFID system can simply peel and stick these tags onto their CD and DVD collection and use the same hardware without any changes.

High Frequency 13.56 MHz.: The High Frequency RFID CD DVD tags is very easy to use provide excellent control of audio and video disks. IC and antenna can also provide these tags as **circular white blank labels** or labels with the **company or institution name and logo** on them for ease in identification. Visible RFID tags can discourage theft as they are a reminder that the disk is secured by radio frequency. Since these tags are actually labels, the RFID vendors can provide them with a paper surface or a longer wearing gloss polyester coating identical to the standard library label line.

Ultra High Frequency Gen 2 860 ~ 960 MHz.: When the requirement is to use **Gen 2 UHF** tags, organization can buy the both **UHF Near Field** short read range tags **or a maximum of about 2 - 3 feet high speed read distance** with a **white gloss polyester Far Field UHF 900 DVD or CD Label**. This distance is attained using a fixed reader and special CD/DVD Tag. The read distance changes with different disk manufacturers as each disk can have a different metallic content which reduces this read distance.

Asset management for retail, libraries, museums, archive backups, virtually any application that uses optical storage disks can use these long read distance RFID UHF tags. Using a fixed reader and having the entrance and exit doors equipped with UHF antenna, any disk going in or out will be read easily, conveniently and quickly to record the event. If required, use a digital I/O board to sound an alarm or lock the exit and entrance doors automatically, or have a pre-recorded message play from the location sound system advising of the potential problem.

16. RFID Tag Characteristics

Difference between powered and non-powered RFID chips (active and passive): Active tags have a battery on the tag. The battery may be used to boost read/write range, allow for larger memories, or add sensory and data logging capabilities, such as temperature sensing. Passive tags receive all of their energy from the read/write device that "powers" the tag to allow it to transmit data.

Memory: Passive tags (non-battery) typically have anywhere from 64 bits to 1 kilobyte of non-volatile memory. Active tags, such as those used in military tags, have memories as high as 128 kilobytes.

The information storage: The majority of passive tags use EEPROM memory. Some are laser programmed at the silicon level. Many active tags utilize battery-backed SRAM.

Passive RFID tag accommodates all the data desired: It depends on the amount of data required. Some passive tags can store up to 1 kilobyte of data on the tag.

The relationship between frequency and tag/antenna distance: Passive high-frequency (HF, typically 13 MHz) and low-frequency (LF, around 125 kHz) systems typically exhibit a read range of less than 3 feet. With HF and LF tag systems, the size of a tag (and hence area of antenna) will have a significant impact on read range. Some applications limit the read range to around 6 to 8 inches. Some newer technologies (UHF systems) do have a longer read range that can be 20 to 25 feet, but these systems are intended for pallets and shipping crates. Read range depends on many factors, but the size of the transponder's antenna, the size of the reader's antenna and its output power are the main factors to consider. With battery-less transponders, long read range and small size features are mutually exclusive.

Availability of smallest tags/labels: The smallest smart label form factors are about 1 inch by 1 inch (25 mm square). Specialized RFID tags have been available in sizes as small as 2 mm by 2 mm.

Printer/encoder encounters a bad tag: If a printer/ encoder encounters a bad tag or fails to verify its data, it prints a void error message over the entire face of the smart label. The printer/encoder automatically attempts to program the next tag. Users can select the number of times they want to try to program a good tag. The printer/ encoder can then be configured to send an error message to the host.

Durability: The typical operating temperature for an RFID inlay (tag) found in most smart labels is between -25° C and 70° C. Storage temperature typically is between -40° C and 85° C. These values will vary from manufacturer to manufacturer and will depend on the tag's components. There are industrial tags available in the market that will withstand temperatures as high as 250° C, which could, for example, stand up to heat sterilization requirements for medical items.

High-pressure pasteurization: This would depend on the tag's construction. Most molded tags can withstand fairly high pressure.

X-rays and tags: For the most part, X-rays will not affect a tag. It depends on the intensity of radiation. Gamma radiation, typically used in sterilization applications, is known to erase or destroy most silicon-based electronic circuits.

17. RFID System Design and Prototyping

The RFID system design and prototyping services involve:

- Engineered design of RFID data collection systems in any or multiple RFID frequencies including ISO/IEC 11784/85 (Low Frequency 125 kHz., and 134.2 kHz. - FDX-A, FDX-B, HDXS), ISO/IEC 14443A & B & ISO/IEC 15693, MiFare, Legic, (High Frequency 13.56 MHz.), all ISO 18000 standards, (Ultra High Frequency 860

MHz. ~ 960 MHz.) UCODE EPC Gen 2, UCODE HSL, UCODE G2XM and others

* System implementation cost comparisons between different frequencies such as LF versus UHF, etc.
* System roll-out road-map
* System prototyping and debugging
* Database design for real time data collection
* Custom tag design and manufacturing
* Program design for local, web based and mobile solutions

RFID PROVISIONING DEVICES

Tag provisioning is also called tag commissioning. RFID provisioning devices set up the RFID tag with a unique serial number so that the tag can be used as a unique identifier. The provisioning device may be a RFID writing device or a printer that can write to a tag. The provisioning device encodes information on the chip within the transponder. These devices typically read the tag (after the tag is provisioned/commissioned) to verify that the tag received the correct information. This post-provisioning verification acts as a quality control. An RFID-enabled printer encoder may be thought of as a provisioning device. It becomes the point at which a tag may have information written to it.

RFID INTERROGATING DEVICES

RFID interrogating devices are called readers or may be reader/writers. A reader is an electronic device used to communicate with RFID tags. A reader has one or more antennas, which transmit radio waves and receive signals back from the tag. A reader may be called an interrogator because it "interrogates" the tag. RFID readers may be either fixed (stationary) or mobile. Fixed readers are typically affixed to stationary positions such as dock doors, truck yard entrances and toll road toll booths. Mobile readers are typically smaller than fixed readers and are movable. Mobile readers may be carried around warehouse by a worker (handheld devices) or may be attached to a forklift.

RFID readers, RFID writers and RFID printers are effectively the base connection points used for an RFID system. Often only a RFID read function may be needed. This simplifies the requirements for the item, but at other times both RFID write and read will be needed. RFID technology allows considerable flexibility in the placement of the RFID readers or RFID transceivers with respect to the RFID tag as line of sight is not a requirement. Barcode systems require a line of sight path to the bar code, they must be orientated so that the barcode reader can see and read the barcode. RFID readers can be located in a position sufficiently close to the item to allow successful reading. This provides significantly greater flexibility and reliability. The RFID reader or reader writer (RFID transceiver) is obviously a key element in any RFID system. It enables the tags to be detected and the data read. While there is not the same level of constraint on the size, they must nevertheless be made relatively compact. In some circumstances there may be environmental requirements, especially if they are to be operated outside or on vehicles, the RFID reader design can pose some design challenges.

RFID READER ELEMENTS

The RFID reader can be broken down into a number of major elements or sub-systems. In addition to these elements, there may also be software.

Antenna : The antenna is an integral element within the RFID reader. The antenna must obviously be tuned to the frequency of operation. It must also be mechanically incorporated into the overall packaging or case design for the RFID reader. For lower frequencies, the RFID reader antenna may comprise a coil, whereas for higher frequencies it may be a form of dipole element. With RFID up to 30 MHz typically using inductive coupling this means that most antennas would comprise a coil, but above this where radiative systems are used and the wavelengths are much shorter, then forms of dipole offer good performance.

Controller : The controller is the area within the RFID reader that provides the control for the system. It will enable the read or read-write processes to be action-ed correctly and any protocols that need to be observed will be initiated in this area. The complexity of the controllers for RFID readers varies considerably dependent upon the application and the system. Some are particularly simple, whereas others will be considerably more complicated.

Network interface : Once information has been gained from a tag, the RFID reader needs to ensure that the required actions are taken. To achieve this, the RFID reader will normally need to communicate with a central controller. Traditionally RS232 or RS422 interfaces have been used, but now there is much greater use of Ethernet, or wireless systems including Wi-Fi, Bluetooth, or Zigbee.

RFID PRINTERS

RFID printers are used in conjunction with RFID smart labels. The RFID printers are devices that have the ability to print the label and also programme or write the tag. There is a wide variety in performance of RFID printers. Some are intended for low volumes RFID label / tag production while others are needed for much higher volumes. Those RFID printers used for low volume label production may require the labels to be applied to the product manually. Others used for high volume production may also have an applicator built into the system.

RFID INTERROGATING ANTENNAS

In RFID applications, antennas consist of two general types, namely the RFID tag antenna and the RFID reader antenna (this RFID antenna may also be part of an RFID transceiver or RFID printer, etc). Inductive antennas are coils that use magnetic coupling; they are most commonly encountered in low-frequency and high-frequency tags and readers, but are also used in ultra-high-frequency near-field tags. Most UHF and microwave tags use variants of the dipole antenna, and are radiatively coupled to the reader antenna, which is often a patch antenna or (less frequently) a Yagi-Uda type antenna. RFID antennas form a particularly important element of any overall RFID system. When designing an RFID antenna system, the frequency of operation, and the coupling mechanism are two obvious points. RFID interrogating antennas send a signal to the RFID tag and receive signals back from tags in the antenna's vicinity. An antenna "transmits" when it radiates an electromagnetic field in response to an applied alternating current. An antenna "receives" when it induces an alternating current in response to an ambient electromagnetic field. An antenna converts radio frequency electrical currents into electromagnetic waves and vice versa. It is a conductor designed to translate an alternating current from an electromagnetic field or to translate an electromagnetic field from an alternating current.

Antennas may be embedded as part of a RFID reader/interrogator or they may be physically separated from a reader. Readers and their physically separate antennas are typically connected with one another via cables or via wiring. Reader antennas come in many form factors. As part of the design of the RFID antenna, parameters such as the radiation resistance, bandwidth, efficiency and Q all need to be considered, so that the resulting design for the RFID antenna meets the requirements and allows the required level of performance to be achieved.

1. Basics of Antenna

As with any antenna, and RFID antenna follows the basic rules of any antenna system. The antenna is basically a form of tuned circuit. Power is fed into the antenna and much of it is radiated. As all passive antennas are perform in an equivalent manner in reception as they do in transmission, it is often easier to look at them as a radiating element as it is often easier to look at the concepts in terms of radiation.

There are a number of useful parameters and definitions for antennas described as below:

Radiation resistance : The amount of resistance required to dissipate any power that is radiated.

Resistive losses : The losses that occur as a result of the resistance of the antenna elements. These losses as well as the radiated power equate to the total input power are significant.

Bandwidth : The band over which the antenna will operate satisfactorily. Normally antennas operate as resonant elements and therefore their performance falls either side of the centre frequency. This must be accounted for in the design of the RFID antenna, or any other antenna for that matter.

Feed impedance : The current and voltage will vary along with the length of the antenna element. Voltage rises towards the ends and the current falls and is also dependent upon the length of the antenna, etc. As impedance is the ratio of current and voltage, this means that the feed impedance varies. To ensure the maximum power transfer the source and load impedances must match, and therefore the feed impedance of the antenna is particularly important to ensure efficient operation.

2. RFID antenna bandwidth

With antennas, very small in comparison to wavelength, the bandwidth of the antenna tends to be small. This can create problems when used with modulation systems that may use wider bandwidths. Some RFID systems send data at very low data rates and in a straightforward manner - others use a subcarrier and this increases the bandwidth required. Keeping in view of the small size of the antenna with respect to the wavelength, the bandwidth of the antenna may be sufficiently narrow. It may lead to the centre frequency may radiate well. The sidebands arising from the subcarrier and modulation may be outside the antenna bandwidth and may not radiate effectively.

The RFID antenna bandwidth can be determined from the following formula:

Bandwidth = F_{tol} + F_{sc} + Max data rate

Where:

Bandwidth is in Hertz.

F_{tol} is the frequency tolerance of the antenna.

F_{sc} is the frequency of the sub carrier in Hertz.

Max data rate is expressed in bits per second.

3. RFID antenna coupling

Many RFID antennas will use near field effects, i.e. inductive coupling. When RFID antennas use these near field effects, the RFID reader and RFID tag will be relatively closely coupled to each other. This has a major effect on the bandwidth and performance of the antenna. Accordingly the RFID antenna coupling must be taken into account when designing the antennas. The RFID antenna is an essential item within any RFID system. It enables the signals to be radiated or received and as a result, its performance is a crucial to that of the overall system. Accordingly the design of an RFID antenna must be carefully considered and its performance optimized to ensure that the whole system can perform satisfactorily.

RF SIGNAL

RF signals are driven by two factors:

1) Frequency
2) Air Interface Protocol

RFID systems can use a variety of frequencies to communicate as there is no single frequency solution. Radio waves work and act differently at different frequencies, a frequency for a specific RFID system is often dependant on its application. High frequency RFID systems (850 MHz to 950 MHz and 2.4 GHz to 2.5 GHz) offer transmission ranges of more than 90 feet, although wavelengths in the 2.4 GHz range are absorbed by water, which includes the human body, and therefore has limitations.

There are a huge range of frequencies currently in use in the RFID world. There are number of factors which affect the choice. A matter of distance comes first in picture. It is found that the greater the distance organization/user wishes to cover, the higher the frequency at which organization/user will operate. What's the difference? Well UHF (Ultra High Frequency) can be read at greater distances than HF (High Frequency). This feature makes it ideal for reading the entire stock of the library at a single pass. Unfortunately that also means it might be as easy to do exactly that when organization/ user only wants to read a few items at a time - like issuing books for example. UHF doesn't support the new data standard for library RFID - ISO 28560.

For example, in libraries of Europe and UK, all HF solutions currently use 13.56 MHz. It means that all of the tags in all of the libraries operate at the same frequency. What's more they all currently use the same "air interface". It enables a reading device to correctly interpret the data when it arrives. Each of the frequency bands has advantages and disadvantages for operation. The lower frequencies 125-134 kHz and 13.56 MHz work much better near water or humans than do the higher frequency tags.

RFID operates in several frequency bands. The exact frequency is controlled by the Radio Regulatory Body in each country. The generic frequencies for RFID are:

* 125 - 134 kHz
* 13.56 MHz
* UHF (400 - 960 MHz)
* 2.45 GHz
* 5.8 GHz

Comparing passive tags, the lower frequencies usually have less range, and they have a slower data transfer rate. The higher frequency ranges have more regulatory controls and differences from country to country. The various bodies that control frequencies include the FCC in the USA and CEPT/ETSI in Europe. Below are some typical frequencies used in RFID signals:

LOW FREQUENCY (LF)

Frequencies: Less than 135 KHz

Typical application: Animal tracking

Features/Limitations: Longer antenna required, shorter read ranges, least susceptible to performance degradations for substances such as metal or liquids

High Frequency (HF)

Frequencies: 13.56 MHz

Typical application: Smart cards

Features/limitations: Shorter read ranges

Ultra High Frequency (UHF)

Frequencies: 433 MHz, 868-915 MHz

Typical application: Supply chain

Features/Limitations: Higher data transfer rates, less issues with multiple tags

Microwave

Frequencies: 2.45 GHz, 5.8GHz

Typical Application: Toll Roads

Features/Limitations: Absorbed by liquids, metal materials issues

1. Frequency Ranges

Since RFID systems generate and radiate electromagnetic waves, they are justifiably classified as radio systems. The function of other radio services must under no circumstances be disrupted or impaired by the operation of RFID systems. RFID systems should not interfere with nearby radio and television, mobile radio services (police, security services, and industry), marine and aeronautical radio services and mobile telephones. The need to exercise care with regard to other radio services significantly restricts the range of suitable operating frequencies available to an RFID system. For this reason, it is usually only possible to use frequency ranges that have been reserved specifically for industrial, scientific or medical applications or for short range devices. These are the frequencies classified worldwide as ISM frequency ranges (Industrial-Scientific-Medical) or SRD (Short Range Device) frequency ranges, and they can also be used for RFID applications.

2. Radio Regulation

The standards drawn up by the ETSI (European Telecommunications Standards Institute) serve to provide the national telecommunications authorities with a basis for the creation of national regulations for the administration of radio and telecommunications.

3. RFID Frequency Bands

There are a number of RFID frequencies, or RFID

frequency bands that systems may use. The frequency used by the RFID system determines many of the characteristics about the way in which it will operate. As a result, determining the correct RFID frequency band is an important early decision in the development process.

RFID frequency band allocations : There is total of four different RFID frequency bands or RFID frequencies that are used around the globe. These are placed widely different areas within the radio frequency spectrum. It enables RFID to choose frequencies that will enable the right system parameters to be obtained.

RFID Frequency Band	Frequency Band Description	Typical Range	Typical RFID Applications
125-134.2 kHz and 140-148.5 kHz	Low frequency	Up to ~ 1/2 metre	Can be used globally without a license. Often used for vehicle identification. Sometimes referred to as LowFID.
6.765 - 6.795 MHz	Medium frequency		Inductive coupling
13.553 - 13.567 MHz	High FrequencyHF Often called 13.56 MHz	Up to ~ 1 metre	Used for electronic ticketing, contactless payment, access control, garment tracking, etc
26.957 - 27.283 MHz	Medium frequency	Up to ~ 1 metre	Inductive coupling only, and used for special applications.
433 MHz	UHF		Backscatter coupling, used for remote car keys in Europe
858 - 930 MHz	Ultra High FrequencyUHF	1 to 10 metres	This band cannot be accessed globally and there are significant restrictions on its use.

			When it is used, it is often used for asset management, container tracking, baggage tracking, work in progress tracking, etc. and often in conjunction with Wi-Fi systems.
2.400 - 2.483 GHz	SHF		Backscatter coupling, but only available in USA / Canada
2.446 - 2.454GHz	SHF	3 metres upwards	Used for long range tracking and with active tags, RFID and AVI (Automatic Vehicle Identification). Uses backscatter coupling
5.725 - 5.875 GHz	SHF		Backscatter coupling. Not widely used for RFID.

Source: RFID Frequencies and Frequency Bands available at http://www.radio-electronics.com/

858 - 930 MHz UHF RFID Band: It is necessary to consider the frequencies that are to be used as spectrum allocations and general regulations vary from country to country. There are different restrictions in different countries for access. The details are given as following:

Country	Comments
North America	Here the UHF RFID band can be used unlicensed within the limits of 915 MHz ± 15MHz (i.e. 902 - 928 MHz). There are restrictions on transmission power.
Europe (less exclusions)	Within this region, the RFID frequencies (and other low-power radio applications) specified ETSI recommendations EN 300 220 and EN 302 208, and ERO recommendation 70 03. These allow RFID operation within the band 865-868 MHz,

	but with some involved restrictions. RFID readers must to monitor a channel before transmitting - "Listen Before Talk".
France	The North American standard is not accepted within France as it interferes with frequencies allocated to the military.
China and Japan	There are no licence free RFID bands or frequencies. However it is possible to request a licence for UHF RFID which is granted in a site basis.
Australia & New Zealand	Within this area the RFID band exists between 918-926 MHz as these frequencies are unlicensed, but there are restrictions on the transmission power.

Source: RFID Frequencies and Frequency Bands available at http://www.radio-electronics.com/

4. Air Interface Protocols

Electronic devices such as RFID readers and RFID tags have to have a common protocol under which they communicate. One may think of a protocol as a language, or rules, by which the devices communicate. Just as in everyday conversations between humans, people select the language in which they choose to communicate English, French, etc. The language protocol provides a platform so that both parties to the communication have a common, agreed to way in which they communicate. Similarly, electronic devices such as tags and readers have to have a common protocol, or set of rules, in order to communicate. Air interface protocols are typically standards-based and thus are identified by the organization that holds the technical standard. In RFID technology, many of the air interface protocols are based on international standards managed by ISO. Here are a few of the common air interface protocols associated with RFID technology:

* **ISO 14443:** Air interface protocol for payment systems and contactless smart cards

* **ISO 15963:** Air interface protocol for vicinity cards
* **ISO 18000-1:** Generic parameters for air interfaces for globally accepted frequencies
* **ISO 18000-2:** Air interface for 135 KHz
* **ISO 18000-3:** Air interface for 13.56 MHz
* **ISO 18000-4:** Air interface for 2.45 GHz
* **ISO 18000-5:** Air interface for 5.8 GHz
* **ISO 18000-6:** Air interface for 860 MHz to 930 MHz
* **ISO 18000-6C:** Air interface for 860 MHz to 930 MHz (more commonly known as EPCglobal Gen 2 or as Gen2
* **ISO 18000-7:** Air interface at 433.92 MHz

5. RFID Frequencies and Transmission Power

The following chart shows the standard radio frequencies ranges used for RFID, what the frequency is used for and the maximum field strength/transmission power. It provides detailed information on UHF and EPCglobal standards by Country.

In Europe, there are numerous different frequencies and different maximum power regulations. Some of the European standards and the USA/Canada regulations have been adopted by other countries. When implementing an RFID system, it is very important to make sure that all of the hardware meets local regulations. In order to understand the local regulations that may or may not apply, organization/user first need to know the standards organizations and how they fit into the local RFID frequency and power transmission regulations.

6. International Standards Organizations

* ISO - International Organization for Standardization
* ITU - International Telecommunication Union
* IEC - International Electrotechnical Commission

Frequency Ranges for RFID Systems

Frequency Range	Description	Fieldstrength/Transmision Power
<135 kHz. (125 kHz. and 134.2 kHz.)	Low Frequency, Inductive Coupling	72 dBìA/m maximum
3.155 ~ 3.400 MHz.	Electronic Article Surveillance (EAS)	13.5 dBìA/m
6.765 ~ 6.795 MHz.	High frequency ISM band (Industrial, Scientific and Medical)	42 dBμA/m
7.400 ~ 8.800 MHz.	High Frequency EAS only	9 dBμA/m
13.553 ~ . 13.567 MHz	High Frequency 13.56 MHz. ISM, inductive coupling, proximity smartcard (ISO 14443, MiFare, Legic, etc) Vicinity smartcard (ISO 15693, Tag-It, I-Code) and item management (ISO 18000-3)	60 dBμA/m
26.957 ~ 27.283 MHz.	High Frequency ISM, inductive coupling, special applications only	42 dBμA/m
433 MHz.	Ultra High Frequency (UHF), ISM, backscatter coupling, active RFID	10 ~ 100 mW
865 ~ 868 MHz.	UHF RFID, Listen Before Talk, backscatter coupling	100 mW ERP Europe
865.6 ~ 867.6 MHz.	UHF RFID, Listen Before Talk, backscatter coupling	2 W ERP (3.8 W EIRP) Europe
865.6 ~ 868 MHz.	UHF Short Range Device (SRD), backscatter coupling	500 mW ERP Europe

Frequency Range	Description	Fieldstrength/Transmision Power
902 ~ 928 MHz.	UHF SRD, backscatter coupling	4 W EIRP. - spread spectrum, USA/ Canada
2.400 ~ 2.483 GHz.	Super High Frequency (SHF) backscatter coupling	4 W - spread spectrum, USA/ Canada
2.446 ~ 2.454 GHz.	SHF RFID and Automatic Vehicle Identification	0.5 W EIRP outdoor, 4 W EIRP indoor
5.725 ~ 5.875 GHz.	SHF ISM, backscatter coupling	4 W USA/Canada, 500 mW Europe

Source: Frequency Ranges. Available at http://www.rfid-handbook.de/

7. National Standards Organizations

* ANSI - American National Standards Institute
* BSI - British Standards Institute

8. Industry Standards Organizations

* AIAG - Automotive Industry Action Group
* EPCglobal - Electronic Product Code global - organization set up to achieve world-wide adoption and standardization of EPC technology and was a joint venture between GS1 and GS1 US, now other countries have also joined (GS1- Canada)
* GS1 - Global organization dedicated to the design and standardization of global standards and solutions to improve the efficiency and visibility of supply and demand chains
* GS1 US - United States version of GS1, there are.more than 100 countries that support the GS1 initiative (GS1-Albania ~ GS1 - Vietnam)

There are three ITU Regions globally:

* Region 1 - Europe, North Africa, Middle East (west of the Persian Gulf), former Soviet Union
* Region 2 - North and South America
* Region 3 - Asia, Middle East (east of the Persian Gulf), Australia and Oceania

While the EPCglobal organization is trying to establish global standards, in many cases the ITU or other groups have already made frequency and transmission power regulations. This makes the EPCglobal standard for UHF multi-frequency and multi-power ratings between the 860 MHz. and 960 MHz. bandwidth using FHSS (Frequency Hopping Spread Spectrum) or LBT (Listen Before Talk).

The following are the current EPCglobal standards for countries or regions starting with North America:

* United Stated and Canada - 902 ~ 928 MHz
* Mexico - 902 ~ 928 MHz

- Puerto Rica - 902 ~ 928 MHz
- Latin America - Costa Rica, Dominican Republic - 902 ~ 928 MHz
- South America - Argentina, Chile, Peru, Uruguay - 902 ~ 928 MHz, Brazil - 902 ~ 907.5 MHz and 915 ~ 928 MHz, Venezuela 922 ~ 928 MHz
- Europe - 865.6 ~ 867.6 MHz. (power varies by country and region)
- North Africa - 862 ~ 870 MHz
- South Africa - 865.6 ~ 867.6 MHz, 915.4 ~ 919 MHz and 919.2 ~ 921 MHz. Different power ratings and technique
- Israel - 915 ~ 917 MHz
- Turkey - 865.6 ~ 867.6 MHz
- Tunisia - 865.6 ~ 867.6 MHz
- Middle East - 862 ~ 870 MHz
- United Arab Emirates - 865.6 ~ 867.6 MHz
- Iran - 865 ~ 868 MHz
- India - 865 ~ 867 MHz
- Thailand - 920 ~ 925 MHz
- Malaysia - 919 ~ 923 MHz
- Philippines 918 ~ 920 MHz
- Malta - 865.6 ~ 867.6 MHz
- Singapore - 866 ~ 869 MHz and 920 ~ 925 MHz - Different power ratings
- Vietnam - 866 ~ 869 MHz and 920 ~ 925 MHz
- China - 840.5 ~ 844.5 MHz and 920.5 ~ 924.5 MHz
- Hong Kong - 866 ~ 868 MHz and 920 ~ 925 MHz
- Taiwan - 922 ~ 928 MHz
- Korea - 910 ~ 914 MHz
- Japan - 952 ~ 954 MHz and 952 ~ 955 MHz - Different power ratings and licenses

* Australia - 920 ~ 926 MHz
* New Zealand - 864 ~ 868 MHz

This list is by no means the full list of all of the different countries that have frequency regulations. To ensure compliance to the local regulations one should make sure that organization/user reviews the current standards prior to purchasing the readers.

RFID BACKEND PROCESSING SYSTEMS

RFID systems have interfaces to corporate business systems such as warehouse management systems (WMS), inventory systems, sales and billing systems, project systems, labor scheduling systems and enterprise resource planning (ERP) systems. Many times a layer of software (often called middleware) will sit between the RFID system and the corporate systems, filtering and smoothing RFID data as well as passing relevant business data on to the corporate systems. Not all RFID data are relevant to corporate enterprise applications. Middleware may be used to place RFID data in context of the business setting and may be used to identify those events that have relevance and require passing on to a company's enterprise applications.

RFID, its application, standardization, and innovation are constantly changing. Its adoption is still relatively new and hence there are many features of the technology that are not well understood by the general populace. Developments in RFID technology continue to yield larger memory capacities, wider reading ranges, and faster processing. It's highly unlikely that the technology will ultimately replace bar code. If some standards commonality is achieved, whereby RFID equipment from different manufacturers can be used interchangeably, the market will very likely grow exponentially.

3

Tools, Techniques and Technologies behind RFID

INVADING AND PERVASIVE!

Radio Frequency Identification (RFID) is evolving as a major technology enabler for tracking goods and assets around the world. It can help hospitals locate expensive equipment more quickly to improve patient care, pharmaceutical companies to reduce counterfeiting and logistics providers to improve the management of moveable assets. It also promises to enable new efficiencies in the supply chain by tracking goods from the point of manufacture through to the retail point of sale (POS). As a result of the potential benefits of RFID, many of the world's major retailers have adopted RFID tagging for pallets and cases shipped into their distribution centres. The consequence of this RFID activity in the retail sector is likely to impact on around 200,000 manufacturers and suppliers globally, and will fuel the market for hardware and software to support RFID.

RFID has many applications outside of the retail

supply chain including some surprisingly familiar ones such as car key-fobs, mass transit, resort lift passes and security badges for access control into buildings. It is often described as a transformational technology in terms of its potential impact on business processes and systems. However, in many ways it is a logical evolutionary step on from the barcode as a way of gaining increased labor productivity through automation. When used in conjunction with allied technologies it can remotely sense objects to determine their identity track their position and detect properties such as pressure and temperature. RFID equipment has steadily fallen in price as volumes increase and microchip unit production costs fall. With the ability to store several k bytes of data in addition to the 'number plate' identifier it could be viewed as a form of 'mass distributed database' that has the potential to become ubiquitous - billions of tags in daily use throughout the world on all objects that are produced, stored, moved, sold and maintained.

TOOLS

1. Understanding Tags

The Tag or Transponder can be either active or passive. It responds to a signal from the Interrogator (reader/ writer/ antenna) which in turn sends a signal to the computer. The Tag comes in a variety of shapes. It is made up from a chip (IC) and an antenna. Depending on the application it may be embedded in glass, or epoxy, or it may be in a label, or a card. The tag can be passive, battery assisted, or active. Passive tags get all their power from the signal sent by the interrogator. As well as using this radio wave to carry the data, the tag is able to convert it into power. This means that the tag is only powered when it is in the beam of the interrogator. The tag then uses a technique called backscatter to reply to the interrogator. This does not involve a transmitter on the tag, but is a means of "reflecting" the carrier wave and putting a signal into that reflection.

Battery assisted tags are just like passive tags (they use backscatter) but they have a battery to provide the power to the chip. This provides a big advantage, because the tag is not dependent on the strength of the carrier from the interrogator to provide the power it needs. As and when it can use all the power from the battery it becomes able to work at a greater distance from the interrogator.

Active tags have a battery, as well as some form of transmitter on the tag. It is concerned with long range. The disadvantage of having a battery is to add cost to the tag, and they run out of power eventually. The tag is made of an IC and an antenna. The IC includes memory and processing capability. The memory may be read only or read/write, the type selected will depend on the application. The tag talks to the interrogator using what is called the air-interface. This is a specification for the interaction to each other. It includes the frequency of the carrier, the bit data rate, the method of encoding and any other parameters that may be significant. ISO 18000 is the standard for the air interface for item management.

Commonly called the anti-collision protocol is also a part of this air interface (if the tag supports it). It allows many tags in the field to talk "at the same time". There are several ways of doing this, and each manufacturer has developed their own way of implementing it.

Two other significant terms are "Reader talks first" (RTF) and "Tag talks first" (TTF). With a RTF system, the tag just sits there, until it hears a request from the interrogator. This means that even though a tag may be illuminated (receiving power) from the interrogator, it does not talk until it is asked a question. With TTF the tag talks as soon as it gets power. In the case of a battery assisted tag or active tag, it talks for short periods of time, all the time. This provides much faster indication of a tag within sight of the interrogator. It also reflects the constant traffic of airwaves. The antenna in a tag is the physical interface for the RF to be received and transmitted. Its construction varies depending on the tag itself and the frequency it operates on. Low frequency tags often use coils of wire, whereas high frequency tags are usually printed with conducting inks. Another form of tag is often called the smart label. This is really a paper (or similar material) label with printing, but also with an RFID tag embedded in it.

Characteristics of RFID Technologies

Characteristics of RFID Technology is given in Table 3.1.

2. Introduction to UHF Tags

Ultra High Frequency (UHF) Tags are becoming more popular for solving difficult RFID problems and meeting business needs across various industries. However questions surrounding High Frequency (HF) and Ultra High Frequency or UHF tags remain. It may be a nice question whether HF or UHF tags are the best solution for meeting business goals.

UHF Tags vs. HF Tags

Till recently HF Tags have been viewed as a more reliable and proven technology. However recent developments established that UHF tags are equally as reliable and often times can offer better performance across many applications. It is necessary to find out how HF tags differ from UHF tags

HF and UHF Tags operate at different frequencies. The maximum range for HF tags is 13.56 MHz and the maximum read range is about 3 feet. Meanwhile UHF Tags operate in the 858 to 930 MHz frequency band and can have a read range of 30 feet (while there are tags with a

Table 3.1. Characteristics of RFID Technologies.

	125kHz	13.56 MHz	433 MHz	915 MHz	2.45 GHz
Active or Passive?	Passive	Passive	Active	Passive (Active also available)	Active and Passive
Designation	Low Frequency (LF)	High Frequency (HF)	Ultra High Frequency (UHF)	Ultra High Frequancy (UHF)	Microwave
Advantages	These are not affected by liquids. Only technology that can work around metal.	Larger data capacity and rate than LF	The Active RFID stands in a class of its own due to long read-range.	Highest read-range for passive tags.	Highly directional, enables precise targeting.
Disadvantages	These are more expensive tags than HF and UHF due to large number of windings necessary. No or poor anti-collision. Low data capacity and rate.	Negatively affected by metal and other materials.	High price of tags.	Highly affected by metals and liquids. Subject to reflection.	Most affected by metals and liquids.
Read Range	Up to 1 m with long-range	Up to 1.5 m with long-range fixed	3 to 100m and even more	Up to about 4 m under perfect	Up to 1 m for passive.

	125kHz	13.56 MHz	433 MHz	915 MHz	2.45 GHz
	fixed reader.	reader.Up to 15 cm with hand-held reader.		conditions with a fixed reader.Up to 1.5 m with hand-held reader.	
Tag Pricing ($USD)	~$0.50 to $5.00	$0.30 to $2.00	$20 and up	$0.20 to $8.00 (highly ruggedized)	
Reader Pricing ($USD)	$50 to $5000 for long-range readers	$200 and up	$600 and up	> $1000 up to $4000 for full portal systems	
Common Uses	Animal Tracking. Industrial Process Tracking. Access Control.	Item tracking. Document/ Library. Smart Cards.	Asset and fleet tracking.	Logistics case/ pallet tracking. Baggage handling.	Logistics. Highway toll collection.
Tag Sizes	From cm-long glass capsules to 1-foot disks.	These are very thin. Postage stamp to credit-card size and larger.	Credit card sized.	Very thin. Credit card sized.	
Notes	Most mature technology. Many existing applications.			Europe uses a slightly lower UHF frequency.	

Source: RFID available at—http://www.rfid-handbook.de/rfid/types_of_rfid.html

longer read range, 30ft are an average).

HF Tags are considered more "mature" tags, since they have been around for longer. While UHF tags are dominant technology for supply chain applications. It is believed that due to their "maturity" and a short read range HF tags are a better choice for item level or near-field applications. However, experiences show that UHF Tags are equally as reliable as HF tags and often can offer better performance. UHF tags do not need to be near field tags to be read at a close range. Unlike HF tags, UHF tags can be read both at close and long ranges. In this context, a 30 ft. tag can also be read from an inch away. UHF tags can be optimized for various applications due to their versatile read range, and be processed at various read distances. UHF Tags are usually applied for work-in-process applications including, inventory tracking, anti-counterfeit applications, identification applications and many more. UHF tags are perfect for applications where a "universal" tag is necessary. It is a tag that can be read at both long and short distances.

Additionally, HF tags are not an option where a longer read range is required. Therefore it is important to consider future applications and extensions of RFID uses when planning on implementing an HF RFID System.

Advantages of UHF Tags

There are several advantages to using UHF Tags. The pattern identifies cost and performances are the two most relevant attributes. The speed is also important. The speed at which UHF tags may be read and the distance over which UHF tags may be read are usually higher when compared to HF tags. Additionally, due to mandates to their supplier to use UHF tags for their supply chain tracking, it is believed that UHF tags will be cheaper. The idea is that as UHF tag production volume increase, cost will decrease. Furthermore, it is suggested that as UHF tags become more popular in item level tagging and the production numbers will reach billions, tag prices will plummet.

These factors clear the operational advantages to implementing a UHF RFID system. There are still myths and misconceptions that need to be cleared.

Misconceptions about UHF Tags

While UHF tags can obviously offer a versatile read distance, many believe there are drawbacks to UHF Tags. For a long time it was believed that

1. UHF Tags could not function on/in water or metal
2. UHF Tags could not provide shorter read range
3. UHF Tags are too large and are inappropriate for applications where a smaller footprint is necessary

2. Some myths and solutions

UHF Tags cannot function on or in liquids or metals

It is well proven that HF tags perform reliably and accurately on liquids and metals. However, for a while it was believed that UHF tags cannot function on/in liquids. UHF tags perform equally well (if not better) on both liquids and metals. UHF Gen 2 Tags not only work on liquids they actually work in containers that are filled with liquids. In 2009 a RFID tag manufacturer demonstrated that a UHF Tag can, not only work on containers containing liquids but in liquids as well, by reading a UHF Tag placed inside a Gatorade bottle.

UHF tags not only work on metal, but these tags actually take advantage of the metal background, utilizing it to increase performance. For example, RCD Technology's Sentry Family of Asset tags are all mount on metal UHF Tags, specifically designed for superior performance on metal surfaces. In short, UHF tags are flexible and can be designed to achieve the best performance required on metal and on - or in - liquids. Additionally, Boeing chose to identify aircraft parts with

UHF Tags to streamline maintenance and pre - flight inspection processes. Boeing used a mount - on - metal UHF tag to identify metal objects.

UHF Tags do not provide contained read range, required for item level tracking.

UHF Tags can provide a full spectrum of ranges required, from near field ranges as small as an inch away to 50+ft. Read range is not strictly determined by the frequency. UHF Tag inlays designed to have a shortened read range will be as good (if not better) as HF tags. Antennas can be designed to have a short, long or complete read ranges. Unlike HF Tags, UHF tags allow for flexibility in read range. These tags can be designed to function in near and far field applications.

3. UHF Tags are too large and are inappropriate for applications where a smaller footprint is necessary

UHF Tags do not have size constraints. UHF Tags can be designed based on a footprint. For example, RCD Technology developed a UHF Tag; with a footprint small enough to be placed on servers that have limited space (36.3mm x 10.9mm x 2.8mm). Additionally, medical UHF labels can be found as small as 9 mm in size. UHF tags provide the flexibility to meet customers' needs, from short range tags to tags small enough to fit in the cap of a vial to large tags that require robust performance in the longest of read ranges.

As one can see from this brief, UHF Tags are the right choice for most applications. UHF can offer the same effective uses as HF. UHF tags can be read at a long and short distances, while HF tags are limited to a very short read range. UHF tags are easily customized to meet application specific goals and can be read on/in water, on metal, and can vary in size. Additionally UHF tags have been implemented in many supply chains and business in many difficult applications.

GENERAL TECHNICAL INFORMATION

Radio Frequency Identification works on the basis of radio-frequency communication between an RFID Tag and RFID Reader. RFID Tags come in two varieties: Passive tags derive power from the reader itself, while Active tags carry their own, often replaceable, battery. Active tags are significantly more expensive, but have a much longer read-range than Passive tags. From the point of view of the Automatic Identification and Data Collection (AIDC) industry, the functionality of RFID technology partially overlaps that of traditional barcode technology. The following table compares some of the advantages and disadvantages of each technology.

Beyond the Passive/Active distinction, RFID technology operates in several different frequency ranges, each with specific advantages and disadvantages. The following table lists the five frequency ranges commonly used for RFID in North America, along with their characteristics.

COMPARISON OF BARCODE AND RFID TECHNOLOGIES

Table 3.2. Comparison between RFID and Barcode Technology Source:

Barcode Technology	RFID Technology
Requires line-of-sight. This is generally considered a disadvantage; however, it can actually be an advantage where precise targeting is required. An operator sees and knows exactly which barcode he is scanning.	No line-of-sight required. This tends to be an advantage for systems that aim to eliminate human intervention, e.g. baggage handling and document inventory systems. RFID has an advantage in dirty or harsh environments where a barcode would become obscured.
Barcodes are scanned one at a time.	Depending on technology used, up to several hundred RFID tags

	can be scanned in a single second by one reader.
Once printed, a barcode cannot be modified.	Some (not all) RFID tags allow information to be modified. Typical capability is 100,000 write operations with a 10 years data-retention lifespan.
Data on barcode could be encrypted; however, they provide no protection from being copied.	RFID tags allow more sophisticated forms of data protection.
Typically cheaper than RFID tags.	More expensive than barcode. The cost of tags can be the one thing that makes or breaks the economic case for RFID. Tags range from $USD 0.15 for huge volumes of disposable tags up to around $USD 5.00 for environmentally protected passive tags in low volumes. Active tags start at around $USD 20.00 each.
Other than the obvious requirement of the barcode being physically visible, the technology is not subject to many variables.	RFID implementation need to consider a number of variables, including electrical and RF noise, proximity of metals and liquids, and environmental conditions like snow and humidity.

OPERATING PRINCIPLES OF RFID SYSTEM: RFID COUPLING

There is a huge variety of different operating principles for RFID systems. The way in which the RFID tag and Reader-writer communicate is known as the RFID coupling mechanism. Terms including RFID backscatter coupling or RFID backscattering, RFID capacitive coupling, RFID capacitive coupling and the like all refer to the coupling method, i.e. the way in which the tag communicates its information to the reader-writer. The form of coupling that is used depends upon the intended application, and in turn the type of RFID coupling used will affect the choice of frequency for the system. There are several ways in which the RFID reader-writer can communicate with the RFID tag. The most important principles - 'inductive coupling' and 'backscatter coupling' are described further.The main RFID coupling techniques that are involved are:

* RFID backscatter coupling
* RFID capacitive coupling
* RFID inductive coupling

The type of coupling used affects several aspects of the RFID system including the range, frequencies needed and other elements of the RFID hardware. The range of the RFID system can broadly be categorised into three:

* Close range - within 1 centimetre
* Remote - between 1 cm and 1 metre
* Long range - more than 1 metre

Of these types of RFID coupling, magnetic and capacitive types are normally used for close range links, inductive coupling for remote links and RFID backscatter coupling for long range links.

1. RFID Inductive Coupling technique

An inductively coupled transponder comprises of an electronic data carrying device, usually a single microchip and a large area coil that functions as an antenna. Inductively coupled transponders are almost always operated passively. This means that all the energy needed for the operation of the microchip has to be provided by the reader. For this purpose, the reader's antenna coil generates a strong, high frequency electro-magnetic field, which penetrates the cross-section of the coil area and the area around the coil. Because the wavelength of the frequency range used (< 135 kHz: 2400 m, 13.56 MHz: 22.1 m) is several times greater than the distance between the reader's antenna and the transponder, the electromagnetic field may be treated as a simple magnetic alternating field with regard to the distance between transponder and antenna.

A small part of the emitted field penetrates the antenna coil of the transponder, which is some distance away from the coil of the reader. By induction, a voltage is generated in the transponder's antenna coil. This voltage is rectified and serves as the power supply for the data carrying device (microchip). A capacitor is connected in parallel with the reader's antenna coil, the capacitance of which is selected such that it combines with the coil inductance of the antenna coil to form a parallel resonant circuit, with a resonant frequency that corresponds with the transmission frequency of the reader. Very high currents are generated in the antenna coil of the reader by resonance step-up in the parallel resonant circuit, which can be used to generate the required field strengths for the operation of the remote transponder. The antenna coil of the transponder and the capacitor to form a resonant circuit tuned to the transmission frequency of the reader. The voltage at the transponder coil reaches a maximum due to resonance step-up in the parallel resonant circuit.

RFID inductive coupling is used for what are termed "vicinity coupled" cards. RFID inductive coupling is defined in ISO 15693 standard, although not all RFID inductively coupled tags need conform to this standard. In terms of operation, inductive coupling is the transfer of energy from one circuit to another via the mutual inductance between the two circuits. For RFID inductive coupling to be used, both the tag and the reader will have induction or "antenna" coils. When the tag is placed close enough to the reader the field from the reader coil will couple to the coil from the tag. A voltage will be induced in the tag that will be rectified and used to power the tag circuitry.

To enable data to be passed from the tag to the

reader, the tag circuitry changes the load on its coil and this can be detected by the reader as a result of the mutual coupling. RFID inductive coupling is a near field effect. Accordingly the distance between the coils must be kept within the range of the effect - normally this is taken to be about 0.15 wavelength of the frequency in use. RFID inductive coupling is normally used on the lower RFID frequencies - often LF, i.e. below 135 kHz or at 13.56 MHz.

The choice of the best form of RFID coupling will depend upon the intended application. Capacitive RFID coupling is used for very short ranges, inductive RFID coupling for slightly longer ranges and RFID backscatter coupling or RFID backscattering is normally used where longer distances are needed.

As described above, inductively coupled systems are based upon a transformer-type coupling between the primary coil in the reader and the secondary coil in the transponder. This is true when the distance between the coils does not exceed 0.16??, so that the transponder is located in the near field of the transmitter antenna.

If a resonant transponder (i.e. the self-resonant frequency of the transponder corresponds with the transmission frequency of the reader) is placed within the magnetic alternating field of the reader's antenna, then this draws energy from the magnetic field. This additional power consumption can be measured as voltage drop at the internal resistance in the reader antennae through the supply current to the reader's antenna. The switching on and off of a load resistance at the transponder's antenna therefore effects voltage changes at the reader's antenna and thus has the effect of an amplitude modulation of the antenna voltage by the remote transponder. If the switching on and off of the load resistor is controlled by data, then this data can be transferred from the transponder to the reader. This type of data transfer is called load modulation.

To reclaim the data in the reader, the voltage measured at the reader's antenna is rectified. This represents the demodulation of an amplitude modulated signal. In the terminology of radio technology the new elementary frequency is called a subcarrier. Data transfer is by the (amplitude-shift-keying) ASK, (frequency-shift-keying) FSK or (phase-shift-keying) PSK modulation of the subcarrier in time with the data flow. This represents an amplitude modulation of the subcarrier.

2. RFID backscatter coupling technique

RFID backscatter coupling or RFID backscattering uses the RF power transmitter by the tag reader to energize the tag. Essentially they "reflect" back some of the power transmitted by the reader, but change some of the properties, and in this way sends back information to the reader. Using RFID backscatter or RFID backscattering, some tags achieve their data transmission by changing the properties of the tags themselves, while others switch a load resistor in and out of the antenna circuit.

RFID backscatter coupling operates outside the near field region, and the radio signal propagates away from the RFID reader. When the signal reaches the RFID tag, this interacts with the ingoing signal and some energy is reflected back towards the RFID reader. The way in which the signal is reflected back depends upon the properties of the tag (or any other object for that matter). Factors such as the cross sectional area, and the antenna properties etc. within the tag all have an effect. In particular the antenna will pick-up and re-radiate energy, and the way this energy is re-radiated is dependent upon the antenna properties - by changing factors such as adding or subtracting a load resistor across the antenna, the re-radiated signal properties can be changed. Over short ranges, the amount of power reaching the tag from the reader is sufficient to allow operation of small low current circuits within the tag. This can be used to drive an electronic switch, e.g. a FET that can switch an antenna load resistor in and out of circuit. This will effectively modulate the returned signal and allow data to be passed back to the reader. In order to allow transmission and reception of a signal at the same time, a directional coupler is often used to allow the received signal to be separated from the transmitted on. Additionally the reader must be able to detect the modulation in the presence of a host of other reflections, although these will normally be stable and not modulated in any way.

Electromagnetic waves are reflected by objects with dimensions greater than around half the wavelength of the wave. The efficiency with which an object reflects electromagnetic waves is described by its reflection cross-section. Objects that are in resonance with the wave front that hits them, as is the case for antenna at the appropriate frequency for example, have a particularly large reflection cross-section.

To pick up such signals it needs neither special probes nor expensive equipment. It requires connection of the ground-clip of oscilloscope-probe with the probe-pin. This makes the probe a 1-turn-sense-coil! The inductivity and enclosed area is enough to pick enough energy from the field to display the signals at the scope, even at lower frequencies around 125 kHz. Next, the probe is placed as close as possible to the transponder and the reader antenna.

3. RFID capacitive coupling technique

RFID capacitive coupling is used for short ranges where a form of RFID close coupling is needed. As the name implies the system uses a capacitive effects to provide the coupling between the tag and the reader. The system is often used for smart cards for which the standard ISO 10536 may be applied. RFID capacitive coupling operates best when items like smart cards are inserted into a reader - in this way the card is in very close proximity to the reader. Rather than having coils or antennas, capacitive coupling uses electrodes - the plates of the capacitor to provide the required coupling.

The capacitance between the reader and card tag provides a capacitor through which a signal can be transmitted, although an earth return is required. The AC signal generated by the reader is picked up and rectified within the RFID tag and used to power the devices within the tag. Again the data is retuned to the RFID reader by modulating the load.

BACKSCATTER RADIO LINKS

Passive and semipassive RFID tags do not use a radio transmitter; instead, they use modulation of the reflected power from the tag antenna. Reflection of radio waves from an object has been a subject of active study since the development of radar began in the 1930s, and the use of backscattered radio for communications since Harry Stockman's work in 1949. The induced current is no different from the current on the transmitting antenna that started things out in the first place: it leads to radiation. (A principle of electromagnetic theory almost always valid in the ordinary world, the principle of reciprocity, says that any structure that receives a wave can also transmit a wave.) The radiated wave can make its way back to the transmitting antenna, induce a voltage, and therefore, produce a signal that can be detected: a backscattered signal. On the other hand, if instead a load that permits little current to flow (that is, a load with large impedance) is placed between the antenna and ground, it seems reasonable that little or no induced current will result. These are currents on the antenna that lead to radiation, there will be no backscattered signal in this case. Therefore, the signal on the transmitting antenna is sensitive to the load connected to the receiving antenna.

To construct a practical communications link using

this scheme, we can attach a transistor as the antenna load. The transistor gate contact is held at the appropriate potential to turn the transistor on current travels readily through the channel. When the gate is turned off, the channel becomes substantially nonconductive. Since the current induced on the antenna, and thus, the backscattered wave received at the reader, depend on the load presented to the antenna, this scheme creates a modulated backscattered wave at the reader. Note that the modulating signal presented to the transistor is a baseband signal at a low frequency of a few hundred kHz at most, even though the reflected signal to the reader may be at 915 MHz. The use of the backscatter link means that the modulation switching circuitry in the tag only needs to operate at modest frequencies comparable to the data, not the carrier frequency, resulting in savings of cost and power.

In order to implement a backscattered scheme, the reader must transmit a signal. In many radio systems, the transmitter turns off when the receiver is trying to acquire a signal; this scheme is known as half-duplex to distinguish it from the case where the transmitter and receiver may operate simultaneously (known as a full-duplex radio). In a passive RFID system, the transmitter does not turn itself off but instead, transmits CW during the time the receiver is listening for the tag signal. RFID radios use specialized components known as circulators or couplers to allow only reflected signals to get to the receiver, which might otherwise be saturated by the huge transmitted signal. However, in a single-antenna system, the transmitted signal from the reader bounces off its own antenna back into the receiver, and the transmitted wave from the antenna bounces off any nearby objects such as desks, tables, people, coffee cups, metal boxes, and all the other junk that real environments are filled with, in addition to the poor little tag antenna.

If two antennas are used (one for transmit and one for receive), there is still typically some signal power that leaks directly from one to the other, as well as the aforesaid spurious reflections from objects in the neighborhood. The total signal at the receiver is the vector sum of all these contributions, most of which are much larger than the wanted tag signal, with appropriate amplitudes and phases, most of which are unpredictable a priori. Thus, the actual effect of a given change in the load on the tag antenna on the receiver signal is completely unpredictable and uncontrollable. For example, modulating the size of the tag antenna current (amplitude modulation) may not result in the same kind of change in the reader signal.

A case where changing the tag reflection from a large amplitude (HI) to a small amplitude (LO) causes the received signal to increase in magnitude without changing phase (the "AM" case). Changing the phase of the tag signal without changing the size of the reflected signal in order to symbolize a local oscillator (LO) state may change the amplitude of the reader signal at constant phase. In order to make a backscatter link work, we need to choose a way to code the data that can be interpreted based only on these changes and not on their direction or on whether they are changes in phase or amplitude. As a consequence, all approaches to coding the tag signal are based on counting the number of changes in tag state in a given time interval, or equivalently on changing the frequency of the tag's state changes.

Therefore, all tag codes are variations of frequency-shift keying (FSK). It is important to note that the frequency being referred to here is not the radio carrier frequency of (say) 900 MHz but the tag (baseband) frequency of perhaps 100 or 200 kHz. A binary '1' might be coded by having the tag flip its state 100 times per millisecond, and a binary '0' might have 50 flips per millisecond. Because the frequency being changed is the frequency at which a carrier is being amplitude modulated, techniques like this are sometimes known as subcarrier modulation. Let's look at one specific example of tag coding, usually known as FM0. In real receivers, noise and interference may be present as well as the desired signal. A certain minimum signal-to-noise ratio (S/N) is necessary for each type of modulation in order that it can be reliably decoded by the receiver. The exact (S/N) threshold depends on how accurate user is trying to be and to a lesser extent on the algorithms used for demodulation/ decoding. For RFID using FM0, (S/N) of around 10 or better (10 dB or more) is usually sufficient.

LINK BUDGETS

To transmit to a tag, a reader uses amplitude modulation to send a series of digital symbols. The symbols are coded to ensure that sufficient power is always being transmitted regardless of the data contained within in. The received signal can be demodulated using a very simple power detection scheme to produce a baseband voltage, which is then decoded by the tag logic. In the corresponding tag-to-reader arrangements, the tag codes the data it wishes to send and then induces changes in the impedance state of the antenna. The reader CW signal bounces off the tag antenna (competing with other reflections) and is demodulated by the reader receiver and then decoded back into the transmitted data.

The reader must power the tag, so far we have avoided coming to grips with the crucial associated question of just how much power the tag needs to get and just how far user can go from the reader and still get it. The amount of power that one needs to deliver to a receiver across a wireless link in order that the transmitted data be successfully received is known as the link budget. Since readers and tags both talk, for an RFID system there are two separate link budgets, one associated with the reader-to-tag communication (the forward link budget) and one with the tag reply to the reader (the reverse link budget).

ELECTROMAGNETIC WAVES, SIGNAL VOLTAGE, AND POWER

1. Electromagnetic Waves

The mass in the universe is composed of dark matter and more recently minted dark energy, about which essentially nothing is known. For those condemned to the world of baryons and leptons, electromagnetic waves are a fact of life. In most textbooks on electromagnetic theory, readers wade through Maxwell's equations and possibly laborious arguments on mysterious exchanges between the electric and magnetic fields launching self supporting structures with little Poynting vectors pointing out of them: all true but unnecessarily obscure.

To expand a bit: every object in the world that has an electric charge creates an electrostatic potential, which falls inversely as the distance. The potential sensed at some distance corresponds to what the charged object was doing at an earlier time (r/c), because signals move at the speed of light $c = 3 \times 10^8$ m/s. However, we almost never care, because electric charge comes in two flavors, positive and negative, and the amount of energy associated with an isolated charge of only one type is enormous: a microgram of hydrogen, split into its constituent protons and electrons and separated by 1 m, could support a mass of 8 million kilograms against the gravitational attraction of the entire earth. So in almost every case, adjacent to each electron with a negative charge is a proton with a positive charge, such that the two cancel, and have no net effect on cellphone conversation. Electric currents similarly give rise to a magnetic vector potential in the direction of the current flow, which again exists everywhere with amplitude decreasing with distance, at a correspondingly delayed time. Most electric currents flow in a balanced loop: the potential from current flowing up cancels that from current flowing down, and again no net effect results on distant observers.

"PAIN POINTS" OF IMPLEMENTING AN RFID SOLUTION

The biggest "pain point" with RFID is its potential to change the existing business processes because ultimately, you can collect much more relevant data and have it in real time. RFID is an enabling technology. None can extract all the benefits of this technology without fundamental business changes, system changes, and data changes. With early adopters, RFID spawns multiple examples of discovering "new ways of doing things."

1. The implications for the network when implementing RFID

The networks that exist today to support bar codes will more than likely be able to support RFID. RFID and bar codes are both technologies that deliver data to a host system; however, there is a main point of difference. Bar codes utilize one-way serialized and periodic data. RFID is two-way. Data passes from the tag to the reader/encoder and then can pass back again, depending on the application or need to update the tag. Data can be delivered from multiple tags effectively in parallel, and—by virtue of not requiring human intervention—can provide more data in real time.

There needs to be bridge software, or middleware, incorporated into the overall architecture to prevent the amount of data that hits the system at the same time from overwhelming it. So RFID requires data filtering and data-flow management, to turn parallel, two-way data into the serialized data that a legacy system can handle. These functions can be also partially handled by the printer/encoders and readers. Another consideration—the need for more bandwidth in the network—depends on how much RFID increases the overall amount of data flow within the network. If existing networks can handle the additional traffic with the speed required by the applications, they should not necessarily need to be upgraded or be any more complex.

2. Obstacles today have been encountered that will not allow use of RFID

Clearly, physics plays a role here. A product may need to be re-engineered with new packaging or a new design in order to achieve optimal performance from an applied RFID system. Or a business process may need to be changed in order to remove obstacles that prevent RFID used in a specific case. There also has been a lot of discussion about liquids and metals inhibiting RFID from working well, but work-around strategies have lessened this as a mitigating factor. At this time, no significant barriers have emerged that would prevent implementing and using RFID technology in a variety of applications.

ELECTRONIC PRODUCT CODES: THE EPC

In 1999, the Uniform Code Council and EAN International teamed with Gillette and Procter & Gamble to fund the Auto-ID Center at the Massachusetts Institute of Technology. The center focused on the cost equation by working with private industry to develop an RFID tag that would be very low cost when manufactured in high volumes. The thought was to enable a 5-cent tag. That way, companies could put tags on more objects and then connect them to the Internet through a network. The Auto-ID center eventual gained the backing of the U.S. Department of Defense and other large global companies, including Kimberly-Clark, Metro, Target, Tesco, Unilever, Wal-Mart. These companies were attracted to RFID because it held promise in offering supply chain visibility—the ability to know the location of any product anywhere in the supply chain at any time.

The 5-cent tag is still an idea - today passive tags cost from 20 to 40 cents, depending on their features, packaging and the volumes purchased. The Auto-ID Center's contribution went beyond trying to create an inexpensive tag. To make an inexpensive tag, many of the features that drove the costs of the tag were removed or greatly reduced. The computing power, memory, communication capability and security features were stripped from the tag. The inexpensive tag became part of a larger networked-base solution. The Auto-ID Center developed the concept of the Electronic Product Code (EPC), a numbering scheme that makes it possible to put a unique serial number on objects. It developed a way for tags and readers to communicate (called an air interface protocol) and designed the concept of a network infrastructure that stores information in secure Internet-based databases. With a networked architecture, a virtually unlimited amount of data associated with a tag's serial number can be stored online, and anyone with access privileges can retrieve it. This method of architecting RFID as a network-based solution created the potential for less expensive tags.

The Auto-ID Center handed off its technology research to a non-profit organization called EPCglobal. EPCglobal has worked to commercialize EPC-based technology, creating a second-generation air interface protocol and developing the network infrastructure - the EPCglobal Network - to enable companies to share data about objects. Here is how it is planned to work. When Company 1 ships a pallet full of product (such as shoes or toothpaste), the tags on the cases and pallet are scanned as the shipment leaves, and software is used to automatically let Company 2 know (via the network) that the shipment has left the warehouse. Company 2 can look up (via the network) data associated with the serial numbers on the shipment and learn what's coming, when it will arrive and so on. When Company 2 receives the shipment, it scans the tags automatically, and a message can be immediately sent to Company 1 to let it know the shipment arrived.

The potential efficiencies created by this kind of visibility for objects hold significant potential. Companies would be able to reduce inventories while ensuring that the right product is in the right place at the right time. And because no humans would have to scan the tags, labor costs and errors would also be greatly reduced.

This EPC concept has the potential to change the way business is conducted. Today, many companies make goods based on a monthly forecast. They then push the goods out into the supply chain and hope they sell. If demand is greater than they forecast, they lose sales. If it is less than forecast, they have excess goods that are sold at a loss or thrown away. It would be much more efficient if goods could be pulled through the supply chain based on real-time demand. RFID readers would monitor how many products are being sold, via retail outlets or via interim steps in the distribution channel. Readers would signal the backroom when the shelves get low and the systems could request more inventories be brought out. When inventory in the backroom gets low, readers there might signal the warehouse to send more products. When inventory in the warehouse gets low, readers would signal the manufacturer to send more products. And so on back through the supply chain to the manufacturer's suppliers. Wal-Mart was the first retailer to require suppliers to put tags on cases and pallets of goods. In June 2003, it told its top 100 suppliers that they should begin putting tags on shipments beginning in January 2005. If the giant retailer's top suppliers began buying tags that would begin to drive the price down. Lower prices would enable more companies to use the technology. Then volumes would increase and tag prices would fall further.

Due to the lack of industry standards regarding the use of personal information that could be encoded on the chips, many privacy advocates have called on companies to state their intended use of the technology. Recently, the Privacy Rights Clearinghouse called on the Federal Trade Commission to regulate the use of RFID. As of yet, however, no federal agency has come forth to regulate the usage of the technology. Technical standardization will bring the cost of RFID down drastically. EPCglobal Inc., which has taken over for the MIT Auto ID Center, is the non-profit organization that is working to standardize electronic product codes and RFID technology.

The EPCglobal Network uses RFID to enable true visibility of objects in the supply chain. The network has five fundamental elements:

1. The **Electronic Product Code (EPC)** is a unique number that identifies an item in the supply chain, whether that be an individual product or a case, or pallet, of many products being shipped.
2. Each silicon chip of each **RFID tag** is encoded with a unique EPC that identifies the product. The silicon chip is wired to an antenna, and, using radio frequency identification technology, each tag communicates to an RFID reader its EPC.
3. The **Object Name Service (ONS)** collects the EPC that is passed on from the reader. The ONS resides on a computer or local application system. It tells the computer systems where to locate information on the network about the Item who's EPC it has just encountered. This information will typically reside on the Internet, making it readily available on a worldwide level.
4. **Physical Markup Language (PML)** is an XML-based language that is used to define data on objects.
5. **Savant** is the middleware technology that coordinates the movement of information over the computer systems.

4
RFID Implementation Issues

HOW IT CAME INTO PICTURE!

Early RFID implementations were fundamentally driven by external mandates. With significant technological improvements, more readily available component options, cost reductions, and shared lessons learned; the technology has proven its value in driving significant operational efficiencies. Industries seek competitive advantages and integrate RFID much earlier into their production processes. Innovative companies are expanding the use of RFID in their supply chain, logistics and asset tracking operations. Industries are gaining improvements in supply chain visibility, reduced out-of-stock situations and reduced counterfeiting.

RFID opens a range of possibilities including smart cards, automated check in, sorting and automated loading of material onto book trucks. These systems require additional planning and financing to implement but vastly increase the productivity improvements of RFID systems. The technology and systems are new and will impact every level of staff in some way. Careful planning and research is required to ensure that implementation runs as smoothly as possible.

Implementation of an RFID system is a complex web of technology, equipment and software that has to be fitted to each organization's needs. This is a brief list of the most important pieces of the RFID puzzle in a library set up:

* ***Tags:*** Every item must be tagged and the tag programmed with the required unique information.
* ***Readers:*** Each service unit needs to have sufficient portable readers to carry out operations.
* ***Check in, check out and sorting:*** The work area systems need to be designed and built. The system allows for the implementation of self-sorting machines and other time and worker health savers.
* ***Self-Serve check out stations:*** One of the prime cost reducers of the system. Implementation of this system requires several decisions around security protocols, fine and bill payment procedures and other policy decisions that will increase or limit the usefulness of these systems.
* ***Software:*** Newer systems come with software that converts the radio frequency data to digital and communicates it to the system. Older versions require third party software to communicate between the two systems.

THE COST OF RFID EQUIPMENT

Few companies are setting goal is to drive adoption of RFID technology to the point where massive numbers of tags are made each year. Costs will decline further as adoption ramps up. RFID tags do not display prices as it is based on volume, the amount of memory on the tag and the packaging of the tag. It can be packed in plastic or embedded in a label. Generally speaking, a 96-bit EPC inlay (chip and antenna mounted on a substrate) costs from 7 to 15 U.S. cents. If the tag is embedded in a thermal transfer label the price rises to 15 cents and up. Low- and high-frequency tags tend to cost a little more.

RFID tags can cost as little as 50 cents or as much as $50 depending on the type of tag, the application and the volume of the order. Active tags (with a battery) can cost far more. When organizations bundle in a sophisticated sensor, the cost can rise to more than $100 per tag. While some tags may cost upwards of $50, the return on the investment may bet ten-fold. For instance, some railroad companies are using $40 tags that are mounted on rail cars. These tags can be reprogramed thousands of times meaning the cost of each time that tag is read and written too is less than one cent. The ROI in that case is tremendous for the rail company.

1. How much does an RFID tag cost?

While discussing about the UHF Readers, most UHF readers cost from $500 to $2,000, depending on the features in the device. Companies may also have to buy each antenna separately, along with cables. Antennae are about $250 and up. The price of readers is expected to fall as companies purchase them in large volumes. Low- and high-frequency readers range in price, depending on different factors. A low-frequency reader model can be under $100, while a fully functional standalone reader can be $750. High-frequency reader modules are typically $200 to $300. A standalone reader can be about $500.

The cost depends on the application, the size of the installation, the type of system and many other factors. In addition to tag and reader costs, companies need to purchase middleware to filter RFID data. They may need to hire a systems integrator and upgrade enterprise applications (warehouse management systems). They may also need to upgrade networks within facilities. They will also need to pay for the installation of the readers. The readers need to be mounted. They need electrical power and to be connected to a network.

2. Cost and calculations: an example of library setup

The costs of an RFID implementation are not low. The tags still cost around one dollar each depending on the capabilities required. Readers are key tools and costly, too. The redesign of circulation areas is additional cost to the system. The labeling and encoding of each item in the library's collection is also a time- consuming and costly step though one time.

3. Calculating ROI for RFID?

It is not easy to calculate ROI from an organization that deployed RFID technology across its supply chain. The cost of RFID technology depends on which method organization chooses to implement in the supply chain. For those who just want to meet RFID compliance, the price tag will be cheaper. Of course, the total cost increases for suppliers who wish to further invest in the technology's capabilities which are cleaning and analyzing all the data. This data will be stored in the RFID chips.

PLANNING

Planning should not leave any gap. There can be stimulating analysis. There are few questions to be asked before implementation:

* What is the use case for RFID in the organization?
* Is the intent to track product movement from supply through distribution?
* Does the application involve locating the right equipment at the right time?
* Do organizations wish to track materials in production and through manufacturing process so organizations can deliver finished products to customers sooner and at a lower defect rate?
* Is the application targeted at containment, addressing market diversion, or even alleviating counterfeits?

1. Site survey as part of the planning phase

A site survey will identify issues related to RF communications and potential electromagnetic interference in the facility, so organizations can design the RFID system accordingly. A site survey will help identify existing in-band RF sources that may present co-existence challenges; such as vintage 915MHz wireless access points and wireless alarms or monitoring systems. A proper site survey is also instrumental in mapping out the antenna RF coverage, power and network architecture.

A site survey is vital to identify equipment requirements and as a vehicle to optimize component placement. Based upon the objectives, organizations can determine the most appropriate programming and read zones. The site survey is also a good point to determine the amount and content of data that organizations need in the tag. Some customers are comfortable with simply a short random number and a link to a central data base. Many customers need a fully programmed ID that has back-up both locally and to the network.

* **Required data:** Programming tags can be accomplished in a variety of ways, but are generally categorized in the following high-level buckets:
  - "slap-'n-ship", where desktop printers/label programmers dispense labels for manual application, typically at the end of the process. This is not the recommended solution, as it introduces significant variability in the tag placement process. This requires some degree of personnel training, and generally does not facilitate RFID visibility until the final stages of the production process.
  - Label applicators placed at the front end of the packaging line provide a more controlled tag placement, which alleviates variability, potentially enhances tag performance, provides

visibility throughout the manufacturing process, and offers scalability.

— Need to be cognizant of the tag commissioning (programming/application) location. Seek locations with inherent case isolation, such as the case erect/fill station, and if correctly deployed, organizations can take advantage of existing equipment, such as reject stations (where the cases are checked for case weight and proper tag ID, otherwise they are rejected).

— Some products may be imposed to further tag commissioning location restrictions if they are subjected to metal detection processes, as is often the case for such items as adhesive bandages.

Once organizations have identified the reader pinch points, they will be better positioned to determine the type, and quantity of RFID components, including applicators, handheld readers, fixed conveyor readers, integrated reader/antenna assemblies, turntable readers, and portals.

2. Choosing the Tag

The use case will help define the tag performance criteria and ultimately drive the tag selection process, but with so many tag options to chose from, the selection process may be daunting. Ultimately, the choice often boils down to:

* cost,
* size and
* performance.

Fortunately, technological and production trends have afforded solutions which fundamentally meet all three requirements. But differentiations remain, though more subtle than in the past. Considerations for passive tags include: Tag price. Tag prices have been reduced very significantly over the past few years, primarily because of technology improvements, production efficiencies, standardization, and most importantly, volume. Previously, tags were often optimized for various material types and SKUs (stock keeping units), which often resulted in a plethora of custom tags. Needless to say, in addition to development and production ramp-up costs, this approach was not economically viable and tag designs were vulnerable to obsolescence, depending upon their particular volumes. But today, with tags are available in smaller footprints, providing better performance and substantially less affected by the materials for which they are applied, the industry has responded in a more narrowly focused tag offering and customers benefit from volume. The tag selection should initiate with a classic, mainstream tag offering, which will allow organizations to reap the benefits of volume production.

Tag size: It is necessary to consider the geometry of the tag that's appropriate for the asset. Different assets require different size and type tags, based on physical restrictions and material properties. As a rule of thumb, smaller tags are less sensitive, resulting in lower read range properties, but for applicable use cases, this is not an impediment and it is often a desired feature. For instance, for item-level, and pharmaceutical applications, tag geometries are often restricted by the product, but by the same token, long range is generally not desired. Today, the majority of the available UHF tags are rather narrow, and typically slightly less than 4 inches long, which provides a viable option for standard 2x4-inch label stock. But a portion of the market requires labels confined to 3 inches in length. In many of these case-level applications, range remains paramount. Newer, more sensitive RFID silicon and efficient tag antenna developments now facilitate these yet smaller, high performance requirements.

Certain applications require tag orientation, Some

tag offerings provide yet smaller geometries (e.g., 2x2 inch) whereby the symmetry facilitates a vertically oriented RF-field (the nominal preference) on narrow height form factors, such as conventional corrugate trays. Don't assume that smaller tag geometries result in lower cost. Consider high volume, general purpose mainstream options first, and then migrate to high-end or custom options if necessary.

Tag performance requirements: Consider the performance requirements, defined by the use case. Some challenging applications, such as tagged vehicles in tollway applications, require very long read distances. Here, larger tags with high sensitivity and high backscatter properties are best suited. Standard high volume tag offerings typically address applications where the requirements may remain somewhat stringent, but are confined to more reasonable read ranges (such as for a dock egress). Beyond read range requirements, consider the material for which the tag is applied. More demanding applications, such as those requiring relatively long read distances of aqueous materials within cases on a pallet, may confine tag selection to only the most sensitive, highest performance tag options.

* ***Fragile Tags*:** This problem appears to be in the process of being solved with new technology that makes for a smooth label with no bumps or other protrusions. The bumps and protrusions of the chip and antenna make the tags vulnerable to damage and vandalism. Older tags still have this problem but several new designs reduce protrusions very well and limit potential damage, in the near future tags may be as thin as paper.
* ***Dual Use Tags*:** Using the tag for both circulation and security means that once the tag system has been beaten the security is also defeated.
* ***Tag Interference Vulnerability*:** Radio waves cannot

be read through metal. Simply wrapping several layers of aluminum foil around an item can defeat the security system. Placing two tags close together can interfere with the signal disabling the read capability of the scanner. Most suppliers have software systems in place to limit this possibility but the interference cannot be removed entirely at this time.

* *Tag Collision*: Readers can become confused when too many signals are received at the same time. System providers have developed complex algorithms to minimize this problem but the vulnerability remains. For certain challenging applications where the material properties have a tendency to alter the resonant tag antenna properties (e.g., potentially detuning the antenna, resulting in a frequency shift and hence lowering performance), a suggestion is to use a world tag with a relatively flat frequency response and broad bandwidth. World tags typically encompass a very wide frequency band, from 860MHz to 960MHz, which generally results in less performance degradation as result of challenging material application. Using world tags is critically important if the product is likely to be read in environments where different reader frequencies will be used. Quite often, world tags are needed for items of export so that guaranteed reads can take place in both the country of origin and at the point of sale. Again, today's mainstream tag offerings address the more common use-cases, including case reads on conveyance systems, and placard tags on pallets.

Label application technique: Pilots often initiate with manual "slap-'n-ship" tag commissioning techniques. This process is generalized by use of a desktop RFID label printer/programmer in which the tag is manually applied to the product. Caution should be exercised with this process. It is true that the initial capital investment is low, but this is often at the expense of non-repetitive, non-optimized and actually high-labor processing costs. It also generally takes place at the end of the packaging process, and offers very little, if any, internal ROI potential through the manufacturing assembly. By contrast, an automated label applicator may result in a higher initial capital investment, but the benefits can be significant, including reduced labor and training, more accurate and repetitive tag placement, which often results in higher performance and more consistent reads. Using an automated label applicator also offers a scalable solution that will likely be required anyway as volumes increase.

Product and packaging characteristics: Properties of the tagged materials and packaging can impact tag performance. It's important to understand the general characteristics of the material under consideration. Aqueous products, such as water-based drinks, liquid detergents, green wood, moist wipes and cosmetics, can be very challenging due to the physical properties of water, which absorbs the available RF energy (like a sponge) and robs it from the tag. Metals and metallized/foil-lined packaging are especially interesting. Tagged properly, the product properties may actually enhance the tag's performance. However, improperly tagged, the results may be dismal.

As if that were not enough, there are those products that have both absorptive and conductive properties. Such applications need special consideration. Spacers are generally used to isolate the tag from the material (as with most products containing metal). A key identifier is the color black. Sometimes carbon or graphite-impregnated plastics, which are typically used for the black coloring, can seriously impede the tag performance. ESD plastic boxes are a prime example (e.g., static resistant materials designed to protect sensitive electronics). Plastic totes used in the electronics industry often have these characteristics.

Near field vs. far field: All antennas emit a magnetic field and an electric field component, but typically one field is emphasized while the other is minimized. Generally speaking, magnetic fields are considered "near field" as their RF radiation field falls off more rapidly (e.g., inversely proportional to the distance cubed) than their "far field" (electric field) counterparts which decay more slowly (e.g., inversely proportional to the distance squared).

Near fields are typically used in applications where short read distances are desired, such as with smart cards or access control where 1 to 4 inch read ranges are acceptable. Magnetic fields are typically associated with inductively coupled loop antennas, which are traditionally not affected by aqueous materials. As a rule of thumb, the read range is limited to approximately 1 wavelength. (At UHF frequencies, 1 wavelength is approximately 13 inches.)

Far fields (aka E-Fields) are typically used for longer range applications, such as reading cases or pallets through a conveyor or dock portal, but they are not precluded from reading tags in very close proximity. E-fields can be easily reduced by attenuating (decreasing) power (much like a dimmer for an incandescent light). Recent technology advancements have demonstrated the feasibility of achieving read distances up to 1 or 2 meters with some standard UHF tags on, or in some cases, in aqueous materials.

Some have been led to believe that inductively coupled loop reader antennas and near-field magnetic loop tag antennas are required to read tags on aqueous materials, or for short, confined read ranges where singulation is required. This can be confusing. Near-field antennas in short range applications are probably best described as "close-coupled" antennas, which means that the electric far-field is simply confined to a very short, controlled read zone. Fortunately for those wishing to minimize their tag variety inventory, tags exist today which couple well to both near-field and far-field signals, and system designs facilitate both "close-coupled" and distant read requirements.

3. Optimizing Tag Placement and Orientation

Tagging is not black magic, but technique helps. Ideal RFID tag placement is not the same as for bar code labels, which are traditionally placed in the lower section of a case. Movement helps alleviate RF "nulls" or dead spots. Much like a cell phone whereby an RF dead spot may result at a particular location, dead spots typically are not as noticeable while the phone is in motion. It's generally easiest to read a tag when there is direct line of sight between the tag and the reader antenna (this is referred to as the "near-side"), but unlike bar codes, line of sight is not necessarily a requirement. RFID performance benefits from air gaps, generally at the top of a typical case, furthest from the conveyor metal, and above aqueous material. If organizations are tagging products with foil or metal, organizations can sometimes take advantage of the reflective or waveguide properties and actually channel the RF "through" the product to enhance the tag's performance.

- Tag orientation also impacts read range. Whenever possible, try to vertically orient dipole tag antennas. Horizontal orientations are prone to miss-reads due to "perpendicular" or "cross"-presentation.
- It should be avoided to place tags symmetrically on cases (for example, avoid the center of the case) to help alleviate "shadowing" or "shading" at the pallet level, which occurs when cases are adjacent to one another. Shadowing occurs when multiple tags are placed very close to one another and the tag antennas "hide" or detune each other, thereby reducing the chance of reading "buried" tags by minimizing their chance to be activated. So in general, tags placed too close to one another can

detrimentally affect performance. But here too, new technological advancements in certain RFID tag silicon and tag antenna designs have greatly alleviated this problem.

4. Configuring the Reader

The EPC Gen 2 protocol offers many opportunities for optimizing performance for any given application. Like a high-end camera, some readers offer predefined configurations (like automatic mode on a camera), and some facilitate fine tuning (like manual mode) for the application. For general applications, it's typically acceptable to use the pre-defined or default configurations. Organizations may wish to fine tune some parameters to enhance performance, but exercise care - without a solid understanding of their contributions, altering some parameters could adversely impact performance. Ensure organizations are well acquainted with the command attributes - advanced training is strongly advisable.

The reader organizations choose can be quite important. While many of the more recent vintage of readers work well, there can be significant performance and / or feature gaps between various makes and models. Don't rely simply on "read rates" for quantitative benchmarking. Gen 2 settings are often optimized to minimize repetitive tag reads, and focus their attention on acquiring unique tag reads. This means that once a tag is read, it is placed in an "accounted" state (known as the B-State). Benchmarking read rates in the own environment will provide organizations with the most accurate measure.

* Filtering tags can also minimize network data traffic and downstream decisions. Use of "added" or "removed" tag filters is an excellent way of making the system more efficient.

* Some readers allow the user to tailor the EPC "Q" (Query) such that it initiates with the anticipated slot count, or is bound to an upper limit.

This can be helpful if the anticipated tag inventory is well defined, but it is also advisable to use the reader's auto-Q adjustment feature if available. This helps alleviate tag contention, resulting in much more efficient tag inventories. Proper "Session" selection can also enhance performance. For larger tag populations, Sessions 2 and 3 may prove beneficial. For lower tag populations where redundant tag reads are required, Session 0 may be best suited, because it facilitates higher repetitive reads.

5. Being Environmentally RF Responsible

Help reduce RF pollution by taking advantage of reader triggering options. Some readers facilitate autonomous modes of operation, facilitating automatic acquisitions only when necessary (vs. continuously on). Use of photoelectric eyes or motion detectors is a good practice, as it reduces potential RF interference between adjacent readers or with nearby in-band appliances. It also helps alleviate inadvertent reads.

During the site survey, look for potential interference sources, such as wireless alarm systems, environmental monitors such as baby monitors, and vintage wireless access points. Readers are generally not impacted by this interference but it's important to make sure that the RFID system is a good citizen and does not adversely affect the performance of other wireless devices. Conduct a baseline test without RF and assess the differential impact with RF enabled.

6. Understanding the System Alterations

RFID vendor offerings are often optimized and certified for compliance as a "system." The user organization must be cautious about deviating from the reader manufacturer's recommended settings or complementary components. Readers are compliant with designated cables, antennas and reader settings. Most systems use 6dBi antennas. dBi units characterize Linearly Polarized antenna "gain." dBic units characterize Circularly Polarized antenna "gain" and take axial ratios into consideration. Note that dBic ratings are typically 3dB higher than dBi ratings.

* Implementing organization should be aware of alternate antennas with higher than the recommended gains.
* Implementing organization should be aware that use of any antenna not explicitly listed as compliant by the reader manufacturer technically violates regulatory compliance and such deviations are the liability of the installer.

Long RF cable extensions can result in power losses, which reduce the available RF transmit power available to the antenna.

* When necessary, use low-loss cables if power losses are not acceptable.

  Where systems are designed to use designated cable lengths (such as 6 meter integrated antenna cables), implementing organization must be cautious about using alternate antennas with shorter cables, as this can increase RF power above compliant levels if the RF power is not adjusted appropriately.

  Implementing organization must be advised that power meters are generally thermally responsive, and the complexities of modulated reader output signals are not easily measured, as the RF output includes various constant wave, modulated signals, and null periods in each inventory cycle, and this often results in an averaged output.

7. Myriad Multipath and Reflections

The multipath signals can interfere with the RF reception. The reception is cancelled due to a "bounced" secondary (reflected) signal that is often out of phase with the primary incident signal. Implementing organization can adjust the tag and antenna position, type, and orientation to improve performance (or adversely affect it).

Metal surfaces can act as mirrors which reflect the incident signals. Out-of-phase reflections can cause nulls as they cancel the incident signal. This effect is generally more pronounced with horizontally oriented tags. A thorough site survey and environmental analysis can help identify troublesome areas. In extreme situations, lowering power, adjusting antenna orientations or adding absorptive material can help rectify the situation.

8. The Antenna Anticipations

Walls, buildings, floors and the surrounding environment will impact the antenna performance, as the antenna serves to focus and direct the RF energy. Common varieties include circularly polarized antennas, linearly polarized antennas, Yagi antennas and close coupled antennas.

Circularly polarized antennas are the most popular, as they are the most accommodating for various tag orientations. These antennas are often used where tag orientations cannot be controlled.

Linearly polarized antennas focus the electric field in one axis. Because of this focused energy, the RF penetration is typically stronger in that axis than that of a circularly polarized antenna. This is an excellent choice for challenging applications but they must be used where tag orientation is controlled. For instance, these antennas are typically used for tollway applications and reading tough cases within a pallet in a manufacturing site.

Yagi antennas, not readily offered by most reader manufacturers, are less common in RFID implementations. These antennas offer a very focused and intense RF beam which generally results in narrower widths, but longer read distances. Yagi antennas typically require professional installation.

Close-coupled" antennas may be required for high density applications where tags are in close proximity and where conventional singulation techniques are not appropriate. These antennas offer low gain and concentrate their energy close to the vicinity of the antenna.

The use case will determine the read requirements and opportunities. A thorough understanding of the process and some creative solutions can often overcome a challenging use case.

RFID AND PRODUCTIVITY: LIBERATION IN LIBRARY

The main benefit of RFID is the increase in productivity. Materials handling is the primary efficiency. Checkout becomes a much faster process as stacks of items can be checked out with one scan of a reader without opening or handling them. This is a dramatic reduction in the steps required during checkout. Fully implemented, the system could do away with a checkout station completely. The patron walks past the exit sensors, the information is read off the smart card right from the patron's pocket and the items are checked out. No contact with staff, no need to interact with a station of any kind. The system also has the potential to use a smart card to increase services to patrons. Information can be embedded in the card including prepaid dollars or other information useful at the checkout or other library service desks. If the installation includes the security addition into the tag then the checkout stations become far more user friendly. All items can be checked out using the self checkout units as no electromagnetic desensitizing occurs that can damage sensitive audiovisual material. Also, the careful placement required by barcode technology is no longer necessary as tag placement is not relevant to the

RFID reader. Inventory, tracking of lost items, shelf reading and all work associated with checking the shelves becomes much easier using the portable readers. Scan a shelf and the reader will read back the missing or out- of-place items either through a report for limited readers or directly off the reader for more advanced versions. Check in becomes far more efficient as entire bins of books can be checked in by moving the bin through a large reader or scanning a truck of books with a portable reader. This process reduces handling of returned material by several steps. Repetitive strain injuries that have plagued libraries will be reduced as many of the repetitive movements at check in and checkout are reduced. If the implementation of RFID is integrated with an automatic materials sorting system, even further reductions in injuries will occur.

PROBLEMS AND CONCERNS: AFTER EFFECTS

1. **Data flooding:** Each tag generating a message each time when passing a reader may be a desired outcome. However, event filtering is required to reduce this data inflow to a meaningful depiction of moving goods passing a threshold. Various concepts have been designed, mainly offered as middleware performing the filtering from noisy and redundant raw data to significant processed data.

2. **Standardization:** The frequencies used for RFID in the USA are currently incompatible with those of Europe or Japan. Furthermore, no emerging standard has yet become as universal as the barcode. To address international trade concerns, it is necessary to utilize a tag that is operational within all of the international frequency domains.

3. **Security concerns:** A primary RFID security concern is the illicit tracking of RFID tags. Tags, which are world-readable, pose a risk to both personal location privacy and corporate/military security. More generally,

privacy organizations have expressed concerns in the context of ongoing efforts to embed electronic product code (EPC) RFID tags in consumer products.

A second class of defense uses cryptography to prevent tag cloning. Some tags use a form of "rolling code" scheme, wherein the tag identifier information changes after each scan, thus reducing the usefulness of observed responses. More sophisticated devices engage in Challenge-response authentications where the tag interacts with the reader. In these protocols, secret tag information is never sent over the insecure communication channel between tag and reader. Rather, the reader issues a challenge to the tag, which responds with a result computed using a cryptographic circuit keyed with some secret value. Such protocols may be based on symmetric or public key cryptography. Cryptographically-enabled tags typically have dramatically higher cost and power requirements than simpler equivalents, and as a result, deployment of these tags is much more limited. This cost/ power limitation has led some manufacturers to implement cryptographic tags using substantially weakened, or proprietary encryption schemes, which do not necessarily resist sophisticated attack. For example, the Exxon-Mobil Speedpass uses a cryptographically-enabled tag manufactured by Texas Instruments, called the Digital Signature Transponder (DST), which incorporates a weak, proprietary encryption scheme to perform a challenge-response protocol for lower cost.

Still other cryptographic protocols attempt to achieve privacy against unauthorized readers, though these protocols are largely in the research stage. One major challenge in securing RFID tags is a shortage of computational resources within the tag. Standard cryptographic techniques require more resources than are available in most low cost RFID devices. RSA Security has patented a prototype device that locally jams RFID signals by interrupting a standard collision avoidance protocol, allowing the user to prevent identification if desired. Various policy measures have also been proposed, such as marking RFID-tagged objects with an industry standard label. RFID security is a very active research field for a few years, with more than 400 scientific papers published since 2002.

4. Passports : In an effort to make passports more secure, several countries have implemented RFID in passports. However, the encryption on UK chips was broken in less than 48 hours. Since that incident, further efforts have allowed researchers to clone passport data while the passport is being mailed to its owner.

5. Shielding: A number of products are available on the market that will allow a concerned carrier of RFID-enabled cards or passports to shield their data. Some people claim that aluminum shielding, essentially creating a Faraday cage, does work. Simply wrapping an RFID card in aluminum foil can make transmission more difficult and is not completely effective at preventing it. Shielding is again a function of the frequency being used. Low-frequency LowFID tags, like those used in implantable devices for humans and pets, are relatively resistant to shielding, though thick metal foil will prevent most reads. High frequency HighFID tags (13.56 MHz — smart cards and access badges) are sensitive to shielding and are difficult to read when within a few centimetres of a metal surface. UHF Ultra-HighFID tags (pallets and cartons) are difficult to read when placed within a few millimetres of a metal surface, although their read range is actually increased when they are spaced 2-4 cm from a metal surface due to positive reinforcement of the reflected wave and the incident wave at the tag. UHFID tags can be successfully shielded from most reads by being placed within an anti-static plastic bag.

CONCENTRATING ON CONTRADICTIONS

One of the primary issues to be concerned with an RFID implementation is the immaturity of the industry. Standards are only just being developed. Tags and readers are being modified and improved at a rapid rate. Competing technologies and non-cross compliant systems still exist. This ongoing development could make equipment and tags obsolete, rendering investments in older equipment useless, and requiring expensive conversions and upgrades. This is especially concerning for multi-unit libraries that may implement one branch at a time over several years. RFID technology will continuously communicate with the library database, the real danger of merging this technology is in letting secure ILS database be contacted/updated/opened several times and thereby make them prone to hackers. Some of the technological problems that may occur with RFID applications in libraries are:

* RFID tags are not hidden since these tags are typically affixed to the inside back cover and are exposed to vandalism
* Some interference from metallic material in book covers, CDs and DVDs does occur and tag reading may not be correct
* RFID tags are easy to shield from readers by ordinary aluminum foils and may not serve the purpose of being anti-theft tools
* RFID reading accuracy declines if many items are read simultaneously (although collision avoidance air interface protocol may be used to minimize this effect)
* RFID reader range depends on its power and antenna size

In addition to these technical problems, there may be other difficulties in implementing RFID systems in libraries. Some of these are:

* Privacy issues
* Forward (future) compatibility and inter-library operations due to differences in tags or software
* Limitations in reorganization of library spaces
* Metal shelves may hinder operations of readers and sensors
* Operational difficulties such as installation of return chutes, provision of power supply near the self check-in/check-out desks or chutes
* Budget restrictions for technology upgradation as RFID tags are more expensive than barcodes and security strips

One Change taking place in today's headlines is the implementation of RFID technology into the consumer market. RFID (Radio Frequency Identification) tags are tiny passive electronic tags the size of the period at the end of this sentence. When scanned, they respond by sending out a radio signal to a distance of about 10 cm containing up to 128 bits of information. The potential they offer is enormous. RFID tags make it possible to reduce the cost of taking inventory by a factor of 100. In addition, they offer a wide range of benefits in security, logistics, anti-counterfeiting and anti-theft devices, warning systems, sensors etc. etc. The list goes on as far as the imagination will take it. Like all technologies, RFID also presents the specter of the double edged sword. RFID tags make it possible to track everything; from the tires on the car, to the magazines someone buys; from the medicine in the pocket, to the clothes on the back. RFID gives the phrase "Tag! You're it!" a whole new meaning. RFIDs and their potential usage stir the imagination. Here's the challenge faced by the RFID industry. How do they promote:

* The Responsible Proliferation of a Technology Beneficial to Society in General?

* When it is so easy to paint this technology with a totally different brush?

* RFIDs will erase the last hope for privacy when everything and everyone is traceable!

GS1 EPCGLOBAL IMPLEMENTATION GUIDELINES

EPCglobal's implementation guidelines aim to streamline particular processes according to standardized practices. One of the goals of the guidelines is to drive EPC adoption by easing the implementation of RFID using existing standards.

1. Returnable Trade Item (RTI/Pallet Tagging) Guideline

This guideline contains detailed descriptions of multi-party business use cases that consider the need for identification and management of pallets and requirements for what data is to be captured in user memory portion of tag, how it is to be maintained as well as security considerations.

2. Conveyance Asset Tag (CAT) Environmental Testing - Technical Implementation Guide

This EPCglobal guidance document provides guidance for use of Passive RFID devices to label and or identify on Conveyance Asset Tags (CAT). The intent of this document is to bring clarity to the level of testing required for RFID devices throughout the transportation industry.

3. Implementation Guide for the use of GS1 EPCglobal Standards in the Consumer Electronics Supply Chain

This Implementation Guide on the use of GS1 EPCglobal standards in the Consumer Electronics supply chain provides an introduction to the CE supply chain, the GS1 organization and GS1 standards specifically concerning EPCglobal. GS1 EPCglobal supports the global adoption of the Electronic Product Code (EPC) as a global standard to enable accurate information and visibility about products in the supply chain. Key implementation procedures indicating where and how EPCglobal standards could be used and the benefits achieved are outlined.

4. Multi-Sourcing EPC Assignment Recommendation

This Multi-Sourcing EPC Assignment recommendation document is meant to address the management of EPC read events using the current EPCglobal Standards in a multi-sourcing environment. This recommendation is intended to aid in avoiding the duplication of serial numbers within the EPC for retailer's EPC assigned objects while identifying the ownership of the registered event.

5. RFID-based Electronic Article Surveillance Guides for Reusable & Disposable Tags

The Strategic Overview Guide and Technical Implementation Guide for GS1 EPCglobal RFID-based Electronic Article Surveillance (EAS) for reusable and disposable tags address all relevant challenges and issues on how to implement an RFID-based EAS solution within a retail environment using existing GS1 and GS1 EPCglobal standards.

5

RFID Solutions for Library Systems: RFID Working in Library

WONDERS IN WORDS' WORLD!

RFID is sparking interest in the library community due to its applications to increase efficiency, productivity and enhance user satisfaction. Globally emerging knowledge-based societies of the twenty first century will need information to sustain their growth and prosperity. In this scenario, who can deny the importance of libraries, which are repositories of knowledge resources? A library stacked with books and other information dissemination processes, has a physical presence. A library is an institution of knowledge acquisition and learning. Current library management systems use barcode technology and security strips. Using barcodes, a library management system can keep records of lending, borrowing and shelving status of items such as books, audio or video tapes, CDs, DVDs, etc. Security strips on library items tag their movements. But barcodes and security strips (electronic article surveillance or EAS) have their limitations. RFID technology can come to the aid of library managers and users because they are slow to read and are prone to sabotaging by thieves. Many libraries are switching over to RFID applications, for example, the Vatican Library. With its priceless, ancient collections of 2 million books and manuscripts, the Vatican Library is now using RFID to track, manage and secure its assets. The main problem these ancient libraries face are thefts, non-returns and mis-filed items. It is expected that by adopting an RFID solution the libraries will be able to control misuse.

We can't restrict the RFID technology just there to tags books and other library assets. It provides a comprehensive route for enhancing all library services and upgrade operations.

RFID SYSTEM COMPONENTS AND THEIR EFFECTS IN LIBRARIES

The adoption of Radio Frequency Identification (RFID) technology by libraries promises a solution that could make it possible to inventory of their collections. In addition, it allows patrons to check out and return library property automatically at any time of the day. Besides speeding up checkouts, keeping collections in better order, and alleviating repetitive strain injuries among librarians, RFID promises to provide a better control on theft, non-returns, and misfiling of a library's assets. This technology is generating ever-increasing interest. RFID system consists of three components: the *tag*, the *reader* and the *application* that makes use of the data the reader reads on the tag.

1. Tag

Also known as a *transponder*, the tag consists of an antenna and silicon chip encapsulated in glass or plastic. The tags contain a very small amount of information. For example, many tags contain only a bar code number and security bit (128 bits) but some tags contain as much as 1,024 bits. Tags range in size from the size of a grain of rice to two inch squares depending on their application. Researchers are now working on tags as small as a speck of dust. Tags can be passive, active or semi-active. An active tag contains some type of power source on the tag, whereas the passive tags rely on the radio signal sent by the reader for power. Most RFID applications today utilize passive tags because they are so much cheaper to manufacture. However, the lack of power poses significant restrictions on the tag's ability to perform computations and communicate with the reader. It must be within range of the reader to function. Semi-active tags are not yet commercially available but will use a battery to run the microchip's circuitry but not to communicate with the reader. Semi-active tags rely on capacitive coupling and carbon ink for the antennas rather than the traditional inductive coupling and silver or aluminum antenna used in passive tags.

Tags operate over a range of frequencies. Passive tags can be low frequency (LF) or high frequency (HF). LF tags operate at 125 KHz, are relatively expensive, and have a low read range (less than 0.5 meters). HF tags operate at 13.56 MHz, have a longer read range (approximately 1 meter) and are less expensive that LF tags. Most library applications use

HF tags: Tags can be Read Only (RO), Write Once Read Many (WORM) or Read Write (RW). RO tags are preprogrammed with a unique number like a serial number (or perhaps eventually an ISBN number). WORM tags are preprogrammed but additional information can be added if space permits. RW tags can be updated dynamically. Sometimes space on the RW tags is locked where permanent data is kept and the rest of the tag is writable. Most library applications use RW tags.

2. Readers

RFID readers or receivers are composed of a radio frequency module, a control unit and an antenna to interrogate electronic tags via radio frequency (RF) communication. Many also include an interface that communicates with an application (such as the library's circulation system). Readers can be hand-held or mounted in strategic locations so as to ensure they are able to read the tags as the tags pass through an "interrogation zone." The interrogation zone is the area within which a reader can read the tag. The size of the interrogation zone varies depending on the type of tag and the power of the reader. Passive tags, with shorter read ranges, tend to operate within a smaller interrogation zone. Most RFID readers in libraries can read tags up to 16 inches away. Readers in library RFID systems are used in the following eight ways.

Conversion station : Where library data is written to the tags

Staff workstation at circulation : Used to check-in and check-out materials

Patron self check-out station : Used to check-out books without staff assistance

Exit sensors : Verify that all books leaving the library have been checked out

Patron self check-in station : Used to check in books without staff assistance

Bookdrop reader : Checks in books when patrons drop them in the bookdrop

Sorter : Automated system for returning books to proper area of library

Portable reader : Hand-held reader for inventorying and verifying that items are shelved correctly.

3. Application

Once the reader reads the tag, the information is passed on to an "application" that makes use of the information. Examples of applications and their uses fall into at least six categories:

1. Access control (keyless entry)
2. Asset tracking (self check-in and self check-out)
3. Asset tagging and identification (inventory and shelving)
4. Authentication (counterfeit prevention)
5. Point-of-sale (POS) (FastTrak)
6. Supply chain management (SCM) (tracking of containers, pallets or individual items from manufacturer to retailer)

RFID is most pervasive in the SCM market. In the SCM market, items are tracked by pallet or container, not by individual item. Once the individual items are removed from the pallet, they are no longer tagged. In contrast, library applications require that each individual item contain a tag that uniquely identifies the item (book, CD, DVD, etc). The tag contains some amount of static data (bar code number, manufacturer ID number) that is permanently affixed to the library item. This information is conveyed, via reader, to the library's security, circulation and inventory applications.

METHODOLOGY

The methodology for implementation in a library can be divided into many phases taking into consideration of budget provision, the types of document holdings, number of volumes, types of items meant for circulation, and the number and types member the institution has. Care should be taken to integrate the library automation package while detailed tender specifications are drawn. Since the technology is new to Indian library environment proper demonstration of the system can be arranged and should visit the library where the system is successfully running. While evaluating the tender the past experience of firm supplying the equipment, tags, reader and software should be thoroughly investigated. The fixing of tags to documents can be initially outsourced then in house arrangement can be done after proper training. The reader should be able to read the other manufacturers RFID Tags. The provision for reading the existing barcode in the document can be made and the required data can be downloaded by interacting with the present database and can be written to the tag. The tags can be over layered with the self adhesive sticker containing the logo of the library or the institution for longer life. Until sufficient confidence is gained with the system, old system in practice can be continued.

Retrospective conversion of already existing stack requires a "programmer" or "conversion station." The conversion of existing barcoded items, including affixing the tags to library materials, takes 15-30 seconds per item depending on the amount of information added to the tag and the skill of the person doing the tagging. Pre-programmed tags, which are used for new acquisitions in libraries that want only identification numbers on the tags, take even less time because they do not involve scanning existing barcodes. The speed of conversion can be increased by dividing responsibility for removing and replacing library materials, converting the barcodes, and inserting the tags among at least three people. It is essential that the tasks be rotated so that no one repeats the same motions over an extended period of time. Almost all libraries tag new acquisitions as part of the cataloging process, however, libraries that have experienced losses of unprocessed library materials from technical services, might consider doing the tagging at the time of receipt in acquisitions. While inadvertent duplicates cannot then be returned, it should significantly reduce losses and facilitate tracking of items in technical services.

Readers : A typical RFID system includes three different kinds of readers, also known as sensors or scanner/wand. These devices designed to detect and read tags to obtain the information stored thereon.

* The types of readers include staff workstations for circulation desk charging and discharging, patron self-charging stations, and longer-range walk-through exit sensors to detect and read an RFID tag passage for purposes of determining whether it is a charged or discharged.
* RFID exit sensors at exits are of two types, one reads the information on the tag(s) going by and communicates that information to a server. The server, after checking against the circulation database, activates an alarm if the material is not properly checked-out. Another type relies on a "theft" byte in the tag that is turned on or off to show that the item has been charged or not. It is then not necessary to communicate with the circulation database. The security system will work even though the online library server is not working.
* The portable scanner or inventory wand, can be moved along the items on the shelves without touching them. The data goes to a storage unit, which can be downloaded at a docking station or a server later on, or it can go to a unit which will transmit it to the server using wireless technology.

Optional Components

Optional RFID system includes the following three components:

RFID label Printer: An RFID-printer is used to print the labels with an individual barcode, library logo etc. When the print is applied, it simultaneously programmed the data in to the chip. After this process, the RFID label is taken from the printer and self-adhered to the book. It also checks each RFID label for function.

Handheld Reader/Inventory Wand: The portable

Handheld Reader or inventory wand can be moved along the items on the shelves without touching them. The data goes to a storage unit, which can be downloaded at a server later on, or it can go to a unit, which will transmit it to the server using wireless technology. The inventory wand will cover three requirements:

* Screen the complete book collection on the shelves for inventory control.
* Search for books, which are mis-shelved.
* Search for individual book requested.

Other applications can be written for the inventory wand, since the system utilizes a personal data terminal (PDT).

External Book Return: Libraries can offer to return books during off hours. External book return is a machine with a slot with a chip RFID reader integrated into the wall. It works the same way as the Self Check -Out Station. The user identifies himself/herself (if required by the library), and then puts the book(s) in to the slot. Upon completing his/her return, the user will receive a receipt showing how many and which books were returned. Since they have already been checked-in, they can go directly back onto the shelves. These units can also be used with sorter and conveyor systems.

Including the Server: The server is the heart of some comprehensive RFID systems. It is the communications gateway among the various components. It receives the information from one or more of the readers and exchanges information with the circulation database. Its software includes the SIP/SIP2 (Session Initiation Protocol), APIs (Applications Programming Interface) NCIP or SLNP necessary to interface it with the integrated library software. The server typically includes a transaction database so that reports can be produced.

RFID STANDARDS IN LIBRARY

The electromagnetic spectrum on which RFID resides is regulated by local governmental bodies. Global standards "define the most efficient platform on which an industry can operate and advance". The International Organization for Standardization (ISO) and EPC Global have been very active in developing RFID standards. The AutoID Center and their commercial offshoot EPCGlobal have also defined specifications and standards. There are two ISO standards pertinent to library RFID systems. The current standard, ISO 15693, was not designed for the item-level tracking done in libraries. Yet, most library RFID tags follow this standard. ISO 15693 was designed for supply chain applications. It defines the physical characteristics, air interface, and communication protocol for RFID cards. Standard designed for item-level tagging will allow for more secure communications between tag and reader.

Library RFID applications must be able to integrate with the library's integrated library systems (ILS). The SIP2 protocol has made it possible for RFID products and library automation products to exchange information. Over the years, shortcomings have been identified in SIP2 and the standard has been diluted as vendors attempt to modify the protocol to suit their needs.

To address the shortcomings of SIP2, the National Information Standards Organization (NISO) convened a "standards development group with the mission of designing a protocol that would encourage interoperability among disparate circulation, interlibrary loan, self-service, and related applications."

ISSUES IN ROUTINES

During a day, library managers have to supervise many activities within their libraries. These activities have to be performed smoothly and simultaneously for the benefit of all persons concerned. Some of the important functions in a library are:

* Check-in/check-out of items
* Shelve items
* Prevent thefts
* Check inventory

If each of these functions is done by conventional methods, they will take time and lead to inefficiencies and unsatisfactory services to the library patrons. Let us see how check-out of items is done with barcodes and EAS. It is a six step process. This is tedious and slow. If by using RFID technology this process is made for time-efficient, then it could be recommended. As far as shelving is concerned, in a conventional library, it is done manually. In most busy libraries, many books, DVDs, audio cassettes and videos must be shelved everyday. This is a time consuming task. Misplacement of items, mis-shelving of books, is quite common. Not only this, repetitive re-shelving task can be health risk to library staff. If automated, all these problems can be avoided. In each and every activity within a library, RFID technology will provide a greater amount of efficiency and error free functioning. It will ensure that there is:

* Quick check-in/check-out of items for convenience of members
* Quick and correct shelving of items
* Complete prevention of thefts
* Quick inventory check

RFID tags contain a memory chip and RF antenna that can send and receive several bits of data. Such RFID tags are known as smart labels or digital identification tags. These tags provide benefits of electronic article surveillance (EAS) as well as barcodes.

KEY FEATURES OF RFID IN LIBRARY

The reliability of the system, its ease of operation, and the flexibility of tagging all kinds of media easily, is important criteria in choosing an RFID system. The main aim for today's libraries to adopt RFID is the need to increase efficiency and reduce cost. Automation and self-service can help libraries of all sizes toward achieving these aims, and RFID has the added advantage that it can also provide security for the range of different media on offer in libraries. The technology can also improve circulation and inventory control, which helps to optimize the allocation of labor and financial resources. This means that libraries can relieve their professional employees of routine work and operational tasks.

All of the tags used in RFID technology for libraries are "passive." The power to read the tags comes from the reader or exit sensor (reader), rather than from a battery within the tag.

A few libraries use "smart" card, which is an RFID card with additional encryption, is an alternative to merely adding an RFID tag on staff and user identification cards. Not only does that identify users for issue and return of library materials, but also for access to restricted areas or services. This would make it possible to make it into a "debit" card, with value added upon pre-payment to the library and value subtracted when a user used a photocopier, printer, or other fee-based device, or wished to pay fines or fees.

Speedy and Easy User self-charging/discharging: The use of RFID reduces the amount of time required to perform circulation operations. For the users, RFID speeds up the borrowing and returns procedures. The Library professionals are released for more productive and interesting duties. Staff is relieved further when readers are installed in book drops.

Reliability: The readers are highly reliable. Several

vendors of RFID library systems claim an almost 100 percent detection rate using RFID tags. Some RFID systems have an interface between the exit sensors and the circulation software to identify the items moving out of the library. Were a library user to run out of the library and not be catched, the library would at least know what had been stolen. If the user card also has an RFID tag, the library will also be able to determine who removed the items without properly charging them. Other RFID systems encode the circulation status on the RFID tag. This is done by designating a bit as the "theft" bit and turning it off at time of charge and on at time of discharge. If the material that has not been properly charged is taken past the exit gate sensors, an immediate alarm is triggered. Another option is to use both the "theft" bit and the online interface to an integrated library system, the first to signal an immediate alarm and the second to identify what has been taken out.

High-speed inventorying : A unique advantage of RFID systems is their ability to scan books on the shelves without tipping them out or removing them. A hand-held inventory reader can be moved rapidly across a shelf of books to read all of the unique identification information. Using wireless technology, it is possible not only to update the inventory, but also to identify items, which are out of proper order.

Automated materials handling : Another application of RFID technology is automated materials handling. This includes conveyor and sorting systems that can move library materials and sort them by category into separate bins or onto separate carts. This significantly reduces the amount of staff time required to ready materials for re-shelving.

Tag life : RFID tags last longer than barcodes because; the technology does not require line-of-sight. Most RFID vendors claim a minimum of 100,000 transactions before a tag may need to be replaced.

Fast Track Circulation Operation : The use of RFID reduces the amount of time required to perform circulation operations. The most significant time savings are that information can be read from RFID tags much faster than from barcodes and that several items in a stack can be read at the same time. The anti-collision algorithm appears to be working well.

DEMERITS OF RFID SYSTEMS

High cost : The major disadvantage of RFID technology is its cost. While the readers and gate sensors used to read the information typically cost around $1,500 to $2,500 each; and the tags cost $.40 to $.75 each.

Accessibility to compromise : It is possible to compromise an RFID system by wrapping the protected material in two to three layers of ordinary household foil to block the radio signal. It is also possible to compromise an RFID system by placing two items against one another so that one tag overlays another. That may cancel out the signals. This requires knowledge of the technology and careful alignment.

Removal of exposed tags : RFID tags are typically affixed to the inside back cover and are exposed for removal. This means that there would be problems when users become more familiar with the role of the tags.

Exit gate sensor (Reader) : While the short-range readers used for circulation charge and discharge and inventorying appear to read the tags 100 percent of the time, the performance of the exit gate sensors is more problematic. They always don't read tags at up to twice the distance of the other readers. There is no library that has done a before and after inventory to determine the loss rate when RFID is used for security.

Invasion of User Privacy : Privacy concerns associated with item-level tagging is another significant barrier to library use of RFID tags. The problem with today's library RFID system is that the tags contain static information that can be relatively easily read by unauthorized tag readers. This allows for privacy issues described as "tracking" and "hotlisting". Tracking refers to the ability to track the movements of a book (or person carrying the book) by "correlating multiple observations of the book's bar code" or RFID tag. Hotlisting refers to the process of building a database of books and their associated tag numbers (the hotlist) and then using an unauthorized reader to determine who is checking out items in the hotlist.

Reader collision : One problem meet with RFID is the signal from one reader can interfere with the signal from another where coverage overlaps. This is called reader collision. One way to avoid the problem is to use a technique called time division multiple access, or TDMA. In simple terms, the readers are instructed to read at different times, rather than both trying to read at the same time. This ensures that they don't interfere with each other. But it means any RFID tag in an area where two readers overlap will be read twice.

Tag collision : Another problem readers have is reading a lot of chips in the same field. Tag clash occurs when more than one chip reflects back a signal at the same time, confusing the reader. Different vendors have developed different systems for having the tags respond to the reader one at a time. Since they can be read in milliseconds, it appears that all the tags are being read simultaneously.

Lack of Standard : The tags used by library RFID vendors are not compatible even when they conform to the same standards because the current standards only seek electronic compatibility between tags and readers. The pattern of encoding information and the software that processes the information differ from vendor to vendor; therefore, a change from one vendor's system to the other would require retagging all items or modifying the software.

BEST PRACTICES GUIDELINES FOR LIBRARIES

As libraries are implementing RFID systems, it is important to develop best practices guidelines to utilize the technology in best way and to keep the privacy concern away. The following may be the best practices guidelines for library RFID use:

* The Library should be open about its use of RFID technology including providing publicly available documents stating the rational for using RFID, objectives of its use and associated policies and procedure and who to contact with questions.
* Signs should be pasted at all facilities using RFID. The signs should inform the public that RFID technology is in use, the types of usage and a statement of protection of privacy and how this technology differs from other information collection methods.
* Only authorized personnel should have access to the RFID system.
* No personal information should be stored on the RFID tag.
* Information describing the tagged item should be encrypted on the tag even if the data is limited to a serial number
* No static information should be contained on the tag (bar code, manufacturer number) that can be read by unauthorised readers.
* All communication between tag and reader should be encrypted via a unique encryption key.
* All RFID readers in the library should be clearly marked.
* ISO 18000 mode-2 tags should be used rather than

ISO 15693.

Given the immature state of RFID implementations in libraries, best practices guidelines are very much in flux because libraries are just now beginning to understand the implementation issues, shortfalls with the technology and the greater privacy concerns. For the library considering RFID, libraries ask potential vendors whether they plan to develop a system that allows for rewriting tags on every checkout and then restoring the ID at check-in. This process eliminates the problems associated with storing static data on the tags and eliminates the problems of tracking and hotlisting. The ability to write new data to the tags during circulation requires the library to use read/ write tags capable of supporting all the check-ins and check-outs the item will require over the lifetime of the item. The library will need to determine how many writes will be required and then identify a vendor who will support such a protocol.

* The institution's obligations with respect to the notice, access, use, disclosure, retention, security and disposal of records;
* A requirement that the institution maintain control of, and responsibility for, the RFID system at all times;
* The designation of a senior staff member to be responsible for the institution's privacy obligation and its policy.

RFID technology promises to change the world. It has the capability of making the personal lives and the work lives in the library more convenient. However, every new technology comes at a cost. In order to remediate those costs, efforts must be undertaken to guide its development and implementation. Libraries should not yet implement RFID systems. Instead, libraries should be among the entities putting pressure on government and industry entities to develop standards, public policy and best practices guidelines for its use. Libraries that choose to implement RFID technologies in advance of policy safeguards being put in place should take extra precautions to follow evolving best practices guidelines. Libraries should continue to protect privacy by ensuring that they are not seen as proponents of RFID before it can be safely deployed. Libraries should work to ensure that RFID products are manufactured and used according to well-established privacy principles. Libraries should refuse to implement potentially unsafe RFID solutions simply because they are convenient. Finally, libraries must be outspoken in their public education efforts related to RFID.

RECENT DEVELOPMENTS

Recent developments in hardware and software for RFID systems have increased the potential of this technology in library automation and security. Today, the one important result for libraries is the ability to use non-proprietary systems, now that the new generation of RFID-chips with standard ISO 15693 (to be integrated into ISO 18000-3) is available. With this technology, libraries do not have to depend on one single supplier for tags. As libraries make a long-term investment, which mainly consists of the quantity of tags needed, this is a very important requirement.

AUTOMATION AND RFID

In a library, an RFID system consists of:

* A smart label
* A reader or a hardware for interrogating the smart label
* A software for controlling the hardware and decoding the responses from smart labels

Since RFID tags do not have line of sight requirements:

* Many items can be read simultaneously even whilst stacked
* Items do not have to be opened and scanned one at a time
* Items in multiple formats like books, CDs, etc. can be read at the same time

Automated counters, which read RFID signals, can help patrons:

* To self check-in/out of items from a height adjustable counter (may be placed anywhere suitable within the library)
* To quickly renew their issued out items
* To obtain a receipt of the transaction

These automated self check-in/out counters are suitable for children and persons with physical disabilities. After a check-out operation, the same RFID system disables the security requirements so that patrons can take the books, CDs, videos, etc. outside the library. Many libraries, where RFID technology is operational, have an external chute, mounted on the exterior walls of the library, for returns of books and other borrowed materials. This is operational 24 x 7. (In principle, the return drop boxes may be placed anywhere - at airports, at town centers, at railway stations, etc.) These return chutes have RFID readers installed in them. The external return chutes can be opened by member's card (secured access). On-board receipt printer can print out the details of the items returned and automatically check items back to the library management system.

Sorting and shelving becomes quick, easy and accurate with RFID technology. Once items are returned, a digital sort conveyor belt can place them in appropriate bins from where staff can place them in their correct shelves. With hand held mobile digital library assistant and inventory wand, the library staff can obtain data about shelving the item and other information immediately (catalogue number, shelf number, etc.). Many libraries have found that with RFID technology inventory and scanning of items:

* Take only 10% or less time as compared to conventional systems
* Misplaced books and other materials can be found easily – the reader can hone-in of misplaced or wrongly shelved items quickly
* Find cataloguing errors and replace incorrect spine labels

Besides these incredible technological advantages, an RFID system in a library can offer the following as well:

* RFID tags are safe for magnetic media such as CDs, DVDs, etc.
* Some RFID tags are rewritable. If, for example, a cataloguing error occurs, it can be rectified quickly
* Less manual handling of items hence better preservation
* RFID 'smart' membership cards can give members: access control in certain areas, make payment (fines or fees) easy, lets them use fee-based library facilities such as the photocopier, internet, etc.
* Staff has more time from routine chores and can therefore provide better service to patrons
* Staff schedules can be made flexible
* Tags last longer than barcodes as reading is contact-less

Solution details : With RFID applications in libraries, all the library assets, namely books, manuscripts, CDs, DVDs, videos, audio cassettes, etc. have to be embedded with RFID tags. The RFID readers and antennas are placed conveniently where library users will have maximum access. In addition there has to be computers that are managing and controlling all the library activities. Thus, in a library, an RFID system consists of:

* A smart label which can be attached to or concealed within a library item - it contains a memory chip and RF antenna to send and receive data and can be programmed
* Hardware for interrogating the smart label
* Software for controlling the hardware, programming the tags and decoding the responses from smart labels in the interrogation zone

The three main system components can be further separated into six more sub-components:

* A smart label or a tag
* A reader
* An encoder or a printer
* A middleware
* A host system or a server
* An application software

Middleware : RFID middleware is a software layer that connects data coming in from tags (on library items such as books, etc.) and readers, to the library management system. Middleware provides a coherent and stable interface between RFID hardware operations and flow of data elements such as membership number, catalogue number, etc. into the library database. RFID middleware solutions provide messaging, routing, and connectivity features required to integrate RFID data into existing library management system.

Server : A server is the heart of an RFID application system. It is the communications gateway among the various components of the system. It receives the information from one or more of the readers and exchanges information with the library databases.

Application software : The applications software would have the APIs (API: Applications Programming

Interface) necessary to interface the RFID system with the server, etc. so as to achieve automated library system.

BENEFITS TO LIBRARY MANAGEMENT

* Uncompromised security within the library
* Efficient collection management system (can be located suitably and made 24x7)
* Uncompromised collection security
* Flexible staff schedules
* Labor saving methods free the staff to help customers
* Higher customer/patron satisfaction levels
* Improved inter-library cooperation
* Better preservation of inventory because of less handling by staff
* Same security and labeling formats for all items such as books, CDs and DVDs, hence better management of databases

BENEFITS FOR LIBRARY STAFF

* Time saving devices free them to help customer better
* Labor saving devices free them from doing repetitive, physically stressful tasks
* Can have flexible working schedules

BENEFITS FOR LIBRARY PATRONS

* Self check-in and self check-out facilities
* Check-in and check-out of all types of items (books, audio tapes, video tapes, CDs, DVDs, etc.) at the same locations
* More staff available for assistance
* Quicker service such as payment of fees, fines, etc.

* Better inter-library facilities, more efficient reservation facilities, etc.
* Faster and accurate re-shelving means patrons can find items where they should be, hence quicker and more satisfying service
* Height adjustable self check-in/out tables are liked by children and physically disabled persons who use the library

MAJOR WORK AREAS

Counter Station : Counter Station is a staff assisted station on services such as loan, return, tagging, sorting and etc. It is loaded with arming/disarming module, tagging module and sorting module. Arming/Disarming module allows EAS (Electronic Article Surveillance) bit inside the tag of the library material to be set/reset so as to trigger/not trigger the alarm of the EAS gate. Checking of EAS status of library material is easy. The staff puts the item on the reader and click on the view to display the information stored inside the tag and status of EAS. There are also feature of Auto Arming and Auto Disarm. Auto Arm/Disarm will automatic arm/disarm library material that is within the Reader range.

Together with circulation module from Library Management System Software, this station is used for the following services:

* Editing and updating of patron's record
* Add and deleting of patron's record
* Generate loan history for a particular patrons
* Managing of fines incurred by the patron
* Arm/Disarm of EAS bit inside the library material
* Program of new library material
* Sort item in accordance to their branch and category number

* The features of this station depend on the module loaded by the Library Management Software.

KEY BENEFITS

* Provide Value Added Customer Service instead of manual daily routine
* Perform media check-in/-out for those patrons who choose not to use the self service system
* Help patron that need assistance from the staff

Shelf Management System : The Shelf Management Solution makes locating and identifying items on the shelves an easy task for librarians. It comprises basically of a portable scanner and a base station. The solution is designed to cover three main requirements:

* Search for individual books requested
* Inventory check of the whole library stock
* Search for books which are miss-helved

All these functions are performed by sweeping the portable scanner across the spines of the books on the shelves to gather their identities. In an inventory check situation, the identities collected are compared with the database and a discrepancy report could be generated. In situations when search function is required, whether for a particular item or an item category, the information is first entered into the portable scanner from the base station, and when a foreign item is found on the shelves, a built-in beeper sound to alert the librarians.

KEY BENEFITS

i. Changes inventory process

* No more book handling: just pass the reader across shelved books to perform an instant inventory.
* Accuracy: book identification numbers are registered in the Shelf-Management Reader. The

data is then downloaded to the central database.

* The fastest inventory 20 books per second.
* Notification: books to be pulled are up-loaded to the reader for quick identification.

ii. User friendly

* Light weight
* Wand allows easy reading of high and low shelves

iii. Saves time and resources:

* Implementers indicate a 75% reduction in human resources required for shelf management activities.

Book-Drop (Return Station) : The Book Drops can be located anywhere, within or outside the library. Possible remote locations outside the library include MRT/train stations, shopping centers, schools, etc. This offers unprecedented flexibility and convenience of returning library items at anytime of the day, even when the library is closed. Patron inserts the library item into the slot. The reader captures the electronic signature and sends to backend system for loan cancellation. Patron is acknowledged by beeping sound and flashlight. Patron's record is updated immediately. This is possible due to the seamless link between Library RFID Management System and the host computer system. As such, users who have reached their loan quota can start borrowing items once they have returned them through the Book Drop.

KEY BENEFITS:

* The ability to return books during off hours.
* Loans for the returned items will be instantaneously cancelled so that patron may immediately borrow again.
* Librarians are able to allocate more time to customer service, as they are free from the labor-intensive loan cancellation activity associated with barcode system.

* Display the return status and printing receipt.
* The design of the Book Drops is such that items cannot be retrieved back once deposited.

Patron Card Process System : Patron card process system is a significant aspect. Components are as following:

* Card Printer (Double Side printing)
* Digital Webcam & Tripod
* Mid Range RFID Reader/Writer
* Capture, Edit & Printing Software

KEY BENEFITS

* Double Side Card Printing sets librarian print patron card without changing side of card.
* Library can make patron card in quick way and without patron to bring their photo.
* It is best way to serve patron.
* Connect to Integrated Library System Directly

Self Check out / in Station : The Patron Self Check-out station is basically a computer with a touch screen and a built-in RFID reader, plus special software for personal identification, book and other media handling and circulation. After identifying the patron with a library ID card, a barcode card, or his personal ID number (PIN), the patron is asked to choose the next action (check-out of one or several books). After choosing check-out , the patron puts the book(s) in front of the screen on the RFID reader and the display will show the book title and its ID number (other optional information can be shown if desired) which have been checked out.

The patron then confirms that he has finished the check-out process and a receipt is printed, showing which books have been borrowed and the return date. The RFID tag in the book is set on quiet as a result no alarm will go off at the EAS gates. It is also possible to use the station for Check-In (return) of books. In this case the patron identifies herself, chooses return and then puts one book or a stack of books onto the reader. She will receive a receipt. If the books were to be taken through the gate now, an alarm would sound. One important point for library productivity is that the whole process is significantly less time consuming than with barcode and magnetic strip system; therefore long lines are avoided and fewer stations are needed for the same process.

KEY BENEFITS

1. To Librarian:
   * Speeds up book check-in / check-out
   * Frees staff to better service patrons
   * Better space planning
   * Increases membership rate
2. To Patrons:
   * Easy to use: books can be read in any orientation
   * Reduces queuing time
   * Provides patron privacy
   * Encourages patrons to come back

Anti-theft Detection : RFID EAS Gates is the anti-theft part of the Library RFID Management System using the same RFID tags embedded in the library items. Each lane is able to track items of about 1 meter and would trigger the alarm system when an un-borrowed item passed through them. The alarm will sound and lights on the gate will flash as patron passes through with the un-borrowed library material.

The EAS Anti-Theft Gates is used to detect RFID tag that is equipped with EAS (Electronic Article Surveillance). It can detect the RFID tags within 1 meter range without interference of magnetic items. Upon detecting of Armed RFID tags, the alarm will sound on the gate. It has option to trigger a Camera to record patrons who trigger the alarm to the Surveillance Station. Theft detection is an integral feature of the chip within the tag. It is a stand-alone technology, which operates independently of the library database.

FEATURES

* Detect EAS Armed RFID tags
* Multi-item detection
* Able to integrate camera with the gate (Option)
* Gale to integrate with Surveillance Station (Option)

KEY BENEFITS

* Single technology is required for both inventory and theft management of the library.
* Library staff is alerted immediately when un-borrowed items pass through the theft detection gates.
* Would-be thieves are deterred by the physical presence of the gates.
* Number of patrons passing through the gates is captured by a counter on the gates Alarm volume can be easily controlled.

FOUR PHASES OF LIBRARY RFID MANAGEMENT SYSTEM

1. Library Security System Only:

It focuses on using RFID Tag replace EM Security Stripe with the aim to detect the Radio Frequency and replace the EM Anti-theft system to avoid the high volume EM field harm human body and without False Alarm. It requires following equipment & parts:

* Anti-theft Detection Gate - Detects the RFID Tag on Book and Check EAS Status, System alarm when the EAS status is ON.
* Security Processing Station - Turns On or Off the EAS Setting of the Chip
* RFID Tags

2. Support Library Circulation

The focus is on using RFID tags replace the barcode in Circulation process. Reading book's access number from RFID than from barcode to borrow and return the book. This aspect targets on no line of sight needed while allowing check-out and check-in several items simultaneously. It concentrates on performing both identification and antitheft in one single operation. It requires following equipment & parts:

* Counter Station
* New Integrated Library System functions - read from RFID and set the EAS status.

3. Patron Self Check-in/Check-out:

The key is to let patron do the check-out and check-in process of books. It provides following:

* Benefit to Librarian:
  - Speeds up book check-in / check-out
  - Frees staff to better service patrons
  - Better space planning
* Benefit to Patrons:
  - Easy to use: books can be read in any orientation
  - Reduces queuing time
  - Provides patron privacy

It requires following equipment & parts:

* Self Check Station - Perform patron self check-out

and check-in process and turn off or turn off the EAS status.

* Book-Drop - Indoor or Outdoor. After patron drop in the book into this station, book's id is checked and turned on the EAS simultaneously.
* Remote Book-drop - Long distance away from library, Acquire better protection of the Book-drop.
* Sorting station - Sorting the return books automotive or manual.

4. Smart and Quick Inventory

The emphasis is on searching and Inventory by reading the RFID using portable reader and portable computer. It creates quick inventory, look up the misplaced material and search. It requires following equipment & parts:

* Portable RFID reader
* Portable Computer - Notebook PC or PDA
* Inventory & Searching software

LIBRARY PROBLEMS ADDRESSED BY RFID

Libraries make use of a bibliographic database to track circulation information about items in a collection. Each book, upon being acquired by the library, is assigned a unique number, usually called a bar code. There is no fixed relation between author, title, and bar code. In today's library RFID deployments, tags are programmed with at least the bar code. In addition, some vendors suggest placing extra information on the tag, such as shelf location, last checked out date, author, and title.

Depending on the vendor, the security check is achieved in at least one of two ways. One method, used by 3M, VTLS, and Libramation among others, stores the status of the book on the tag; a specific bit, often called a "security bit," reveals whether the book is checked in or checked out. It is important to note that the security bit does not necessarily affect whether the tag can be read. The security bit must be correctly set at every check-in and check-out, or else false alarms may be triggered. A second method does not store the circulation status on the tag. Instead, the readers query the bibliographic database for the circulation status of the book as it passes through the exit sensors; this introduces issues of latency due to query time. Privacy concerns in today's deployments have focused on the bibliographic database and short range of RFID readers. Without the bibliographic database, an adversary cannot directly map a bar code number to the title and author of a book, and so cannot immediately learn the reading habits of people scanned. Some library RFID proponents have argued that an adversary without the database and with only short range readers poses little to no risk.

Libraries are suffering from crunching budget. It is difficult for libraries to keep the library staffed and open. RFID is seen as a way to address the staff shortages. Self-check systems reduce the number of staff needed at the circulation desk. Self-check systems have become very popular with both patrons and staff. RFID self-check systems allow patrons to check-in or check-out several, rather than just one, books at a time. For example the San Francisco Public, Berkeley Public, and Santa Clara City Libraries all report that reduction of repetitive stress injury RSI injuries is one of the reasons they have implemented, or are considering implementing, RFID systems. With RFID-enabled tools, inventory-related tasks can be done in a fraction of the time. A whole shelf of books can be read by the reader with one sweep of the portable reader. It reports missing or mis-shelved books. The ability to inventory items without handling them is an additional benefit for libraries and archives handling sensitive materials. Sorting can be accomplished automatically with RFID. As books are dropped into the book drop, the reader reads the tag and uses the automatic sorting system to return the book back to the shelves, the stacks or the hold area.

Security is another aspect of library operations that may be greatly improved with RFID-based security systems. A single tag can be used for identifying items and securing them. As patrons leave the library, the tags are read to ensure that the item has been checked out. Librarians also report that lost or hidden items are more easily retrieved using the portable readers. Many DVDs being hidden by patrons for their own use later, staff located these lost items and returned them to circulation using the RFID reader. But blocking the tag from being read by using foil or by removing the tag puts into question whether RFID is truly the best approach for library security. Currently readers are already being developed that are capable of scrambling the data on tags.

COST OF IMPLEMENTING RFID SYSTEM IN LIBRARIES

The cost of implementing an RFID system for a library with 40,000 items is at $70,000 and $166,000 for a library with 100,000 items approximately. However, these estimates are probably high, given that costs are continuing to fall for items such as tags — now down to $.20-$.75/each from $ .85 and servers (now below $5,000, then estimated at $15,000). The actual cost to any library will depend on which RFID modules the library uses. The individual RFID components required are as follows:

* Exit Sensor
* Portable Scanner
* Self-Checkout Unit
* Bookdrop Unit
* Exit Readers

The most expensive aspect of any RFID system is the cost of the tag and the cost of placing the tag on each item. Tags range from 20 cents to 75 cents for books. The costs go up when tags are placed on other media such as CDs, DVDs and tapes ($1 to $1.50). Customized tags (inclusion of library logo) increase the costs of each tag further. Once a library purchases tags, they are committed to the tags and the vendor. Once a library selects an RFID system, it is unlikely that anything short of catastrophe could motivate a library to spend the money and labor required to physically upgrade the tags." At 75 cents per tag, a 100,000-item library would spend $75,000 on tags alone. A laborer can install three tags per minute (new installation). The cost of programming the new tags, removing the old tags and placing the new tags in each item makes converting or upgrading the library's RFID system highly unlikely.

ROLE OF LIBRARIANS

RFID technology introduces an ethical dilemma for librarians. The technology allows for greatly improved services for patrons especially in the area of self-checkout. It may reduce repetitive stress injuries for library workers. And yet, the technology introduces the threat of hotlisting and tracking library patrons. Libraries have traditionally acted to protect and defend the privacy of their patrons. Some libraries are implementing a technology before proper safeguards have been developed. Library use of RFID technology serves to legitimize the technology in the eyes of the community. Therefore, it is incumbent on the library community to ensure that the technology is developed in concert with established privacy principles. Usually library use of RFID follows best practices guidelines consistent with library values.

1. Problems faced by library

RFID technology will continuously communicate with the library database. The danger of merging this technology is in letting secure the database be contacted/updated/opened several times. It makes them prone to damage. Some of the technological problems that may occur with RFID applications in libraries are:

* RFID tags are easy to shield from readers by ordinary aluminum foils and may not serve the purpose of being anti-theft tools
* RFID tags are not hidden since these tags are typically affixed to the inside back cover and are exposed to vandalism
* RFID reading accuracy declines if many items are read simultaneously. Although collision avoidance air interfaces protocol may be used to minimize this effect.
* Some interference from metallic material in book covers, CDs and DVDs does occur and tag reading may not be correct
* RFID reader range depends on its power and antenna size

In addition to these technical problems, there may be other difficulties in implementing RFID systems in libraries. Some of these are:

* Limitations in reorganization of library spaces
* Privacy issues
* Metal shelves may hinder operations of readers and sensors
* Forward (future) compatibility and inter-library operations due to differences in tags or software
* Budget restrictions for technology upgradation as RFID tags are more expensive than barcodes and security strips
* Operational difficulties such as installation of return chutes, provision of power supply near the self check-in/check-out desks or chutes

2. Multi-Lingual Automated RFID Self Serve Library System

A multi-lingual automated library system can be designed to facilitate the circulation and organization of small to large - sized collections of books, CD's and DVD's and other reading materials. All materials are equipped with an RFID label that provides monitoring and security for the materials available for borrowing. The RFID tag is read each time an item is loaned out or returned to the library. It is done without manually typing in or writing out book code numbers.

Patron or member workstations employ touch screens for ease of legibility and simplicity of use. ISO RFID cards are used to automatically identify authorized patrons. High Frequency 13.56 MHz RFID labels are utilized for the reading and viewing materials which eliminate all data entry errors. To ensure that all books are checked out properly, the RFID label contains a special setting that must be changed by the RFID reader in order to remove the material from the library. Patrons or members simply place their ID card on the RFID reader and then login to a Self Serve Station. They can borrow materials, browse the library, review their history, return materials, reserve an item that is currently on loan, and other member activities.

* If the member does not have their ID card they can alternately enter their ID number using the numeric keypad displayed on the touch screen.
* Members borrow materials by placing them on the RFID reader pad which reads the tag ID allocates the materials to the member and then deactivates the EAS part of the RFID tag.
* To assist in accuracy all materials can have full colored images to show exactly what the material looks like that is being returned or loaned out. At the end of the session the patron is provided with a

receipt for any items borrowed or returned.

* If the materials are not correctly read and checked by the Self Serve Patron Stations then the alarm will sound when the patron tries to leave the library going through the RFID gate system.
* The system registers each borrowing transaction including the return due date for each item based on item and member rules. If the item is not returned before this due date, penalties can be applied to the member.

Members simply slide their returned materials through the access door and the automated RFID reader will read the materials RFID tag and then update the library system and provide the member with a receipt from the internal receipt printer. The automated drop box improves library returns and borrowing as the staff can determine exactly how many items are in the drop box and members can also see the materials that are available in the drop box from the self serve patron stations. In addition the materials are stored in bin-locations (shelves, cabinets, rooms, etc...) so finding materials is very easy. Since libraries can have many rooms, cabinets, etc., the library is also divided into Zones that represent a physical area in the library (floor, hall, room, group of shelves)

Some systems use gloss polyester labels for long RFID label life. These labels can have Library information such as name, logos, etc., printed on it during manufacturing. For CD and DVD cataloging few RFID systems provide a circular disk label with a gloss polyester finish. For organizations that want to move to the current state of the art technology and use PC/SC and CCID compliant for both ISO 14443A and ISO 18092 devices for NFC smart payments, or be able to swap out any reader without changing any programs or setups. Few vendors provide the latest NFC 13.56 Reader Writers. Library system in a variety of 13.56 ISO protocols designed to suit a variety of budgets and user needs are available. ISO 14443A, ISO 15693 and ISO 18000-3 standards are available for both the readers and labels used in the system.

3. Examining the potential of RFID

The experiences of previous applications of "new" technology, bar codes and EM security, are many disappointments in the system solutions. These disappointments are concerned with the direct application of the technology and the systems interaction with the integrated library system (ILS). The exception is being circulation desk applications with a direct coupling to the ILS. Since increased efficiency in operations is the primary reason for adopting new technology in libraries there are a number of questions that must be addressed when examining the potential for RFID based systems in libraries.

Price : Will libraries benefit from industry trends towards cheaper components, or will a low number of library oriented suppliers maintain a high price compared to other applications? The expectations of low price must not be unrealistic, but a general growth in the industry should also benefit library applications.

Longevity : what is the economic life time of an RFID system? Specifically what is the life time of the RFID tags? Is there a limited number of re-writes before the tag must be replaced? Are there climate factors that affect the wear and tear? What ageing tests have been made?

Rigidity : What physical requirements does the RFID system impose on the library operations and flows? What software requirements?

Reliability/Robustness : As bar code systems can be fooled by photocopies, or counterfeit Printouts of the bar code; there can be means to bypass an RFID station unnoticed. Can a metal (or lead) container be used. There is a real danger of a "knock-out device" available to the consumer market as discussed as part of the privacy concerns about RFID in the retailing sector. The RFID tagged consumer items are traceable after sale so opens up possibilities of unwanted (by the consumer) research. The Auto-ID Center at MIT was quoted to work for a "RFID-chips killing device" for a low cost. Potentially big adoptors of RFID technology in retailing applications are sensitive to privacy concerns. How will these developments affect library applications of RFID?

Return on investment : Calculation of tangible cost savings is an uphill task. For a specified library type and size, is difficult to determine the payback time of the investment?

Interoperability/integration : The lack of standards was sometimes quoted as a major reason why RFID technology has not reached its potential adoption level. Standards are emerging but what is the adoption level of today's suppliers of library solutions. The choice of supplier may be an irreversible decision. The systems communication between the RFID system and the ILS may require the use of a proprietary software protocol? What is the cost (and price model) for the license? Can RFID technology replace EM security functions in addition to replacing bar code identification? Is there, in that case, any risk that installed EM devices (labels, rods, etc.) will interfere and disturb the RFID system?

Technology traps : Which factors are responsible for choosing the system solution (or supplier)? Can the RFID tags be used in alternative solutions (from an alternative supplier)? What about the future technological developments to be incorporated.

4. Benefits of Using RFID In Library

* RFID improves library workflow by
* reducing non-value added work processes
* Improves staff productivity
* Improves customer service

* Assist inventory check with ease.

Book Label : Better performance label (Copper-etched antenna assembly) with the customization of surface top material with library logo design, etc. It provides personalization of label to assist library operations, easy book identification for shelving process. It assists the traceability of book allocation and enhances book return processes by full automation of check-in, EAS activation and system updates completed simultaneously in the self-return chute. Last not the least, it allows better accuracy in book collection.

CD / DVD Label : It has better performance label (Copper-etched antenna assembly) with the customization of surface top material with library logo design, etc. it provides the personalization of label to assist library operations and eliminates additional time and processes spend on checking the availability of CD/ DVD in disk covers. It enhances easy CD/DVD identification for storage process and protects CD/DVD collection from theft.

NEW DEVELOPMENTS

1.SMART DBL

SMART DBL is the latest development, which incorporate RFID and Barcode in a Smart Book Label. Additional barcode is provided for libraries that have co-existence of barcode and RFID system running in parallel. It is an all in one solution for libraries that are integrating from barcode to RFID system.

Benefits Of Smart DBL

* Personalization of label to assist library operations Customization of label to include encoding for barcode system
* Reduce additional process and cost for barcode printing and encoding

* Improve productivity

2. SMART X- Range

It minimizes assembly process during CD/DVD label application and blocking of contents on CD/DVD on applied surface. It liberates from pasting of donut CD/DVD label on disk and pasting of overlay on applied surface. It allows no more inserting of CD Booster for better performance. Smart X-Range eliminates the above 3 steps for normal CD/DVD label application. It has better performance / reading distance, as compared to normal CD/DVD label. It minimizes A 3 in 1 solution for better performance CD/DVD label. For example, Singapore Patent No. 200206895-5 and US Patent No. 7,245,219. Europe Patent is currently in process.

Benefits of SMART X- Range

* Personalization of label to assist library operations
* Reduce additional process and cost for conventional
* CD/DVD application
* Improve productivity
* Protect CD/DVD collection from theft

LIBRARY ARCHITECTURES

None of today's library RFID tags employ read passwords or other read access control. Because the identifier on the RFID tag never changes throughout its lifetime, the ability to read the tag at will creates several privacy risks. First, the adversary may determine which library owns the book and infer the origin of the person carrying the book. In particular, bar codes for libraries with the innovative bibliographic database have well-known, geographically unique prefixes. Vendors may also place library IDs on tags to prevent tags from one library from triggering readers at another. Learning origin data can be a privacy problem.

Any static identifier can be used both to track and hotlist books. In book tracking, the adversary tracks a book by correlating multiple observations of the book's RFID tag. The user may know the title and author of the book if the bibliographic database is available. Here the static identifier can still be used to track the book's movements.

1. Role of Collision Avoidance IDs

In particular, many tags use a globally unique and static collision ID as part of their collision avoidance protocol. This typically will allow unauthorized readers to determine the tag's identity merely through its collision-avoidance behavior. EPC Class 1 13.56 MHz tags use their EPC identifier directly in a similar collision-avoidance protocol.

* ISO 18000-3 MODE 2 also specifies a 64-bit manufacturer ID. The ID is not used directly for collision avoidance. The collision avoidance protocol requires the generation of random numbers, however, and the standard specifies the use of "at least a 32-bit feedback shift register or equivalent." While it is not explicitly specified, it is expected that each tag will have a globally unique seed in practice. In particular, 32 bits of the 64 bit manufacturer ID are defined to be a globally unique "specific identifier"; it would be natural to use this specific identifier to seed a PRNG. If a 32-bit LFSR is used, then tags can be uniquely identified.
* In EPC 915 MHz tags, there are three different modes for "singulation" or collision avoidance, one of which uses the globally unique Electronic Product Code (EPC) ID. The choice of modes is controlled by the reader.

An adversarial reader can simply ask the tag to use its EPC ID; because there is no authentication of this command, the tag will obey. As a consequence, any library system using one of these tags will be vulnerable to tracking and hotlisting of books and patrons. The collision-avoidance behavior is hard-coded at such a low layer of the tag that, no matter what higher layers do, privacy will be unachievable. This is unfortunate, because it means that much of today's RFID hardware is simply incompatible with privacy for library patrons. It is also dangerous, as vendors and libraries may implement privacy-enhancing methods that focus on tag data and then be unaware that tags are not in fact private.

2. Write Protection

In deployments with rewritable tags, some method must be used to prevent user from writing to the tag. Otherwise, user can commit acts of vandalism such as erasing tag data, switching two books' RFID data, or changing the security status of tags with "security bits." For example, the EPC 13.56 MHz tag specification, as well as ISO 18000-3 MODE 1, include a "write" and a "lock" command, but no "unlock" command. In addition, write commands are not protected by password; this is consistent with a supply chain application that writes a unique serial number to a tag, and then never needs to re-write the number. The lock command is only an optional part of the ISO 18000-3 MODE 1 standard. In ISO 18000-3 MODE 2, locking is also irrevocable, but protected by a 48-bit password. Few refer to irrevocable locking of the security bit as a security bit denial of service.

It is also possible for an active adversary to bypass the write lock mechanism by racing a legitimate reader. After waiting for the legitimate reader to unlock the tag, the adversary can then send write commands which will be accepted by the tag. In practice, tags may be left unlocked by accident if a tag is prematurely removed from a reader's field of control before the tag can be re-locked.

3. How to manage Tag Password Management

The ISO 18000 standard and EPC specifications only

allow for static passwords sent in the clear from reader to tag. As noted, current deployments do not seem to use read passwords, but write passwords are employed. There are two natural approaches to password management: (1) use a single password per site; or, (2) endow each tag with its own unique password. If a single password is used for all tags, then a compromise of any tag compromises the entire system. In deployments that use writable security bits, the write password is used on every self-checkout; in systems with read passwords, exit sensors must use the read password every time a book leaves the library. In either case, passwords are available to a passive eavesdropper.

One of the problems with simple passwords is that a passive eavesdropper can overhear the password. The channel from tag to reader is much harder to eavesdrop than the channel from reader to tag. A simple protocol for enhancing passwords in RFID tags was proposed as part of the EPCglobal Gen II standards process. The main idea is for the tag to send a random nonce to the reader; an adversary who misses the nonce cannot recover the password from reader to tag communication alone.

6

RFID Solutions for Library Systems: Concept of RFID for Libraries

ROLLING IN!

RFID is the latest technology to be used in library circulation operations and theft detection systems. RFID-based systems move beyond security to become tracking systems that combine security with more efficient tracking of materials throughout the library, including easier and faster charge and discharge, inventorying, and materials handling. This technology helps librarians reduce valuable staff time spent scanning barcodes while checking out and checking in borrowed items. RFID is a combination of radio -frequency-based technology and microchip technology. The information contained on microchips in the tags affixed to library materials is read using radio frequency technology regardless of item orientation or alignment.

HOW RFID CAME IN BETWEEN THE BOOKS!

A Radio Frequency Identification tag is an electronic device that holds data. Typically these tags are attached to an item and contain a serial number or other data associated with that item. In passive RFID technology, the tag carries no power source, but is instead powered by a radio signal from a separate RFID reader. RFID tags operate under severe restrictions compared to most personal computers, or even most embedded systems.

* First, an RFID tag is powered only when within range of a reader. This means that the tag has only an extremely limited amount of time to carry out computation.
* Second, an RFID has extremely few gates, and many of these are taken up by logic required for basic operation.

While some low-end smart cards and tags have incorporated constructions based on stream cipher designs, no standardized low-gate primitive exists. Simple password comparisons and XOR (a binary operation like AND and OR) operations are all that can be expected on most current- generation RFID tags. In addition, an RFID has almost no physical security. The extra transistors might be used to enable cryptographic primitives on tags of equal cost as today's tags. It is more likely, however, that economic pressures will lead manufacturers to focus on ever-cheaper tags with a feature set similar to current generation RFIDs. Because tags are manufactured on a massive scale, even a half cent difference matters. For example, the RFID manufacturer Alien Technologies RFID tags used in libraries operate on the 13.56 MHz band and are manufactured by several companies, including Checkpoint Systems, Texas Instruments, and TAGSYS. Checkpoint and TAGSYS make proprietary tags, while the TI Tag-It! platform follows the ISO 15693 standard. ISO tags and TAGSYS tags are then resold by a variety of integrators, including 3M, TechLogic, and VTLS. Checkpoint tags, on the other hand, are installed only by the library services division of Checkpoint.

Recently, a new standard for RFID, ISO 18000, reached final stages of approval. ISO 18000-3 defines the physical interface and commands for 13.56 MHz tags. The 18000-3 standard is further divided into two "MODEs." MODE 1 is intended to be backwards compatible with the command set defined in ISO 15693, but standardizes various elements of the RF interface. MODE 2, on the other hand, is intended to be a next-generation RFID standard capable of supporting high-speed data transfer and communications with large numbers of tags at once. In addition, MODE 2 tags are explicitly required to support a random number generator and a small amount of semi-nonvolatile RAM. While MODE 2 tags are beginning to be manufactured, no library RFID vendor currently offers them for deployment.

The EPCGlobal consortium also publishes a series of specifications for RFID tags. These tags are aimed at supply chain markets and do not have a presence in the library setting. These tags also incorporate a special "kill" command (The kill command is used on Linux and other Unix-like operating systems to terminate processes without having to log out or reboot (i.e., restart) the computer.) that renders the tag permanently inoperative; while the kill command is protected by password, reads and writes are not. The 13.56 MHz tags used by libraries have several material differences from the 915 MHz tags considered for supply chain applications. First, the bandwidth available to 13.56MHz tags is strictly limited by regulations in the US, the EU, and Japan. Second, the read range of 13.56 MHz tags is much less than that of 915MHz tags. As a result, RF air interface protocols, such as collision avoidance, differ between 915 MHz and 13.56 MHz tags.

RFID tags communicate with the reader by passively modulating a radio signal broadcasted by the reader. Because a reader is little more than a radio transceiver, this means that attackers will be able to obtain illegitimate readers that can be used to query RFID tags from some distance. Library RFID vendors claim that their readers can interact with tags from a distance of 2 feet (for large sensors at library exits), and hand-held readers might work up to 8 inches away from the tag. These distances are limited primarily by regulation on reader power and antenna size.

Comparing to the 915 MHz tags (used in supply chain and retail applications), can be read from a distance of eight meters or more. Even a few feet of read range is sufficient for scanning people passing through doorways and other close spaces. In fact, the sensors used to detect theft of library books look have similar read range to, the RF-based anti-theft sensors already used in thousands of shops. One unusual aspect of RFID communication is an asymmetry in signal strength. Tags respond by passively modulating a carrier wave broadcast by the reader. It will be much easier for attackers to eavesdrop on signals from reader to tag than on data from tag to reader.

Many RFIDs may be in range of a reader at the same time so the collision-avoidance protocols must be used. The details of these protocols are often kept secret in proprietary tags. The ISO 18000 standard, however, specifies collision-avoidance protocols for each of its two modes, as does the EPCGlobal suite of tag protocols. These protocols require a separate identifier, (a collision-avoidance ID) that may be independent of the data stored on the tag. The collision-avoidance ID can often be used to track tags.

INITIATION

New technologies have always been of interest for

libraries, both for the potential of increasing the quality of service and for improving efficiency of operations. At the present time, when libraries of all kinds (public, research, special) are facing economic hardships the overwhelming reason for considering new technologies is the potential for cost savings in the operations and the management of material flows. Academic libraries and public libraries face growing challenges in managing the assets of their collection and maintaining or improving service levels to patrons. Over the past twenty years, libraries have grown their services to their patrons from simply offering books and periodicals to now adding multi-media items such as videos, CDs and DVDs. There are also growing concerns on the security of the collection assets and maintaining (or improving) service levels to the patrons. Certain capabilities are listed here:

* RFID tags replace both the EM security strips and Barcode.
* Simplify patron self check-out / check-in.
* Ability to handle material without exception for video and audio tapes.
* Radio Frequency anti-theft detection is innovative and safe.
* High-speed inventory and identify items which are out of proper order.
* Long-term development guarantee when using Open Standard.

Various advantages are as described

* Chip is user-friendly and multi-purpose
    - Media is re-writable
    - More ergonomic for staff
    - Theft control more efficient
    - Stack check out (more than one book at a time)
* Efficient self-check

   - Use fewer resources for that repetitive work
   - No waiting lines
   - 24 hours return (automatic book return)
   - Privacy
- Inventory control
   - Faster
   - Done more frequently
- Shelf order function

DATA MODELS

As RFID solutions begin with a chip - attached to an aerial -being read by a device of some sort. There is need to know the actual data being read and its format? That's where the data model comes in. ISO 28560 defines the elements available to any community wishing to use it. In the UK that community comprises the libraries and their suppliers. They decided to use roughly half of the available elements to build a UK "data model". This performs several tasks. Two of the most important are the definition of a "map" of the data locations on the chip, and a list of the values that may be stored in each one. Outside of the data model a few data elements on the tag are already defined by the RFID industry but the rest of the space can, at the moment, be used in whatever way the user or supplier decide.

In general it is true that most library RFID suppliers have deliberately been very limited in their use of this space - and that's a good thing! Often the only data stored is the barcode number and an owning library identifier. However each supplier uses a different data model to store even these two pieces of information with the result that tags from one supplier cannot readily be used by another. Some libraries and RFID suppliers are adding more data to the chip, to meet local needs or to collect additional information about usage. This created a nightmare for servicing companies (or jobbers) in trying to meet the demands of an increasingly fragmented market. Even libraries using the same RFID Company weren't always able to read each other's data. The lack of a common standard also held back the development of consortia as many members were unable to share their resources as planned.

In this context, ISO 28560 (2010) and will be the first international data model for library RFID. The UK data model was agreed by all the major suppliers to the UK market in 2009 and endorsed by a statement issued by 2CQR, 3M, Axiell, Bibliotheca, DTech, Intellident, and Plescon in a newly proclaimed "RFID Alliance" in October of that year.

BOOK CATALOGUING BY USING RFID

Radio Frequency Identification systems have been in use in libraries for book identification, self-checkout, anti-theft control, inventory control and sorting and conveying of library books. These applications can lead to significant savings in labor costs, enhance customer service and prevent book theft and provide a constant record update of media collections. Use of RFID tags replaces the barcodes in Circulation process. Reading book's access number from RFID to borrow and return the book. RFID tags have been specifically designed to be affixed into library media including books, CDs, DVDs and tapes. It is thin, flexible and thus can be laminated between paper and plastic. With special method to attach to books, patron is totally unaware that the tag is there. The advantages of RFID tags over barcode / magnetic stripe:

* Faster scanning of the data stored on the RFID tag.
* High speed inventory - reducing time of staff for 'shelf-reading' & other inventory activities,
* Automated return of materials that speeds up sorting of materials & re-shelving

* A longer lifecycle.

WORLD WIDE LIBRARY APPLICATIONS

We can understand the applications by going through the number of examples. The New Orleans Public Library had all of its branches damaged by Hurricane Katrina in late 2005. The Alvar branch was completely rebuilt, out of necessity, and was reopened in a few days. When it did, all of its books and CDs/ DVDs had RFID tags, which were used to check books in and out. The RFID equipment, the implementation of tags, and the software integration were jointly donated by Tagsys and ITG.). New Orlean's RFID implementation is small in comparison to Shenzhen's new public library in China. The enormous library - over 500,000 sq ft - had around 2 million books tagged. Seattle's RFID library project is the largest in the world, with Shenzhen's second. One such library is at the University of Guelph, in Ontario, Canada, which has had their books tagged since at least 1991, possibly earlier.

THE BASIC INFORMATION OF RFID SYSTEM FOR LIBRARIES

An RFID system comprises three components; a tag, a reader and an antenna.

1. RFID Tag

The tag is paper thin, flexible and approximately 2"x 2" in size which allows it to be placed inconspicuously on the inside cover of each book in a library's collection. It consists of an etched antenna and a tiny chip which stores vital bibliographic data including a unique ID number to identify each item. This contrasts with a barcode label, which does not store any information, but merely points to a database.

2. RFID Reader & Antenna

These components are available in various shapes and sizes to suit respective applications within the library, and are often integrated into one enclosure for that specific purpose (ie patron self check-out machines, inventory readers). The reader powers the antenna to generate an RF field. When a tag passes through this RF field, the information stored on the chip is decoded by the reader, and sent to the PC or central server which in turn, communicates to the Library Information System.

Key Attributes of RFID

Unlike barcodes, RFID tags can be read;

* Through desktops and book covers (no line of sight is required).
* in any orientation to the reader/antenna
* while moving (i.e. while being deposited through a return chute)
* several at a time
* from distances of several inches to the antenna

RFID improves library workflow, staff productivity and customer service with these attributes. However, the ability to conduct inventory counts without removing a single book from the shelf is what really separates RFID from preceding technologies such as barcodes.

Some Recommendations for Libraries

* The selection of a system to automate library functions such as a circulation or an integrated library system (ILS) launches the library into a *long-term commitment* to that system. If a library selects an RFID-based system, it is likely to retain that system for a long time, primarily because of its high cost as well as the time-consuming structural changes that the library must engage in to successfully use the new system.

* Has the library community examined where the RFID technology is going and how it is expected to evolve? Even though privacy and civil liberties impacts might not be of concern now because RFID systems in libraries are essentially "closed" systems, that may not be the situation forever. For example, the publishing industry is exploring the use of RFID to tag books.
* The library community must monitor and shape the direction vendors and jobbers are taking with RFID. It is important for the library community to resist efforts that move RFID from its present proprietary ("closed") status to that of the RFID standards in the marketplace (the EPC standard developed by MIT's AutoID Center) - or that incorporate the use of marketplace tags with library-specific tags, a situation that could occur if the publishing industry adopts RFID. Also, it is especially critical for the library community to require effective, auditable security of RFID systems.
* The library community must conduct a comprehensive technology assessment of RFID as soon as possible to enable librarians to make the best possible decisions involving the implementation of this technology. Such a risk-benefit analysis would include an investigation of the potential privacy and civil liberties implications and the best methods to mitigate these harms.
* The library community is a leader in protecting intellectual freedom and user privacy. Libraries have an opportunity to be societal leaders and a "model industry" in the way in which they adopt RFID-based circulation and ILS systems.

Best Practices, based primarily on Principles of Fair Information Practices:

Privacy policy : If the library has not adopted a privacy policy, it should develop one that encompasses its implementation of RFID.

Notice / Openness : The library must make public its policies and practices involving the use and maintenance of RFID systems, encompassing tags, readers, and the associated database(s).

* Library users have a right to know that items contain RFID tags and generally where readers are located. They have the right to know the technical specifications of those devices.

  Labeling must be clearly displayed and easily understood.

  — Any tag reading that occurs in the library environment must be transparent to all parties. There should be no tag-reading in secret.

Purpose specification : Library users must be given notice of the purposes for which tags and readers are used.

Collection limitation : The collection of information should be limited to that which is necessary for the stated purpose. There should be no personal data encoded in RFID tags.

Security safeguards : There must be auditable security and integrity in transmission, databases, and system access, including the use of encryption.

Accountability : The library must inform its users who they can contact for questions and complaints.

COMMON SCENARIOS AND PREPARATIONS IN EUROPE

The adoption of RFID technology by libraries promises a solution that could make it possible to inventory hundreds of thousands of items in their collections in days instead of months. In addition, it would allow patrons to check out and return library property automatically at any time of the day. Besides speeding up checkouts, keeping collections in better order, and alleviating repetitive strain injuries among librarians, RFID promises to provide a better control on theft, non-returns, and misfiling of a library's assets.

London Libraries Consortium uses RFID for Self-Service, Stock Management and Extended Opening. Benefits to libraries of RFID includes: Quicker stock taking, frees up librarians from admin-heavy tasks to have more interaction with library users etc. Increase opening hours. Radio Frequency Identification was being extensively used by members of the London Libraries Consortium for self-service, improved stock management and to facilitate longer opening hours. Around a third of London is served by the consortium's library services. The most recent RFID installations are at Tooting (Wandsworth), Wandsworth Town, Clapton (Hackney), John Jackson (Enfield), Enfield Town and Gants Hill (Redbridge) where public library technology specialist Axiell has provided Bibliotheca Venus machines which offer easy-to-use self issue and return facilities in over 1,000 libraries worldwide. Wandsworth libraries has had RFID in its Earlsfield library since 2008 and at the new Wandsworth Town library since it opened in August 2009. Wandsworth Town library is in a listed building and comprises a series of rooms rather than an open plan library. In order to provide maximum opening hours over six days a week and to make best use of staff, it was decided to use self-service machines so that we did not need to have a librarian permanently in each of the six rooms. People are wowed by being able to put a whole pile of books on the machine and the machine reading the information from them all at once. Tooting Library serves a strong Asian community and the RFID machines have been set up so that the instructions are available in the relevant language such as Urdu or Punjabi, 87% of issues at the branch are through RFID machines.

The technology will be introduced to the remaining Wandsworth libraries by the end of the year.

In the east of Greater London, Havering carried out stock taking using hand held RFID readers for the first time last year. Six of the borough's libraries have RFID, the latest being the new Central Library in Romford, the remaining four libraries gain the technology in 2010 with a target of 80% of transactions to go through the machines last year. At neighboring borough Redbridge, the technology is also used for stock management as well as self-service. Gants Hill library offers extended opening hours thanks to the introduction of RFID. Staff has been trained in the new customer service model and actively encourages patrons to use the RFID machines and 86% of issues are now going through self-service.

Brent Libraries had 10 fully self-service libraries with the remaining two facilities to follow suit by autumn 2010. There the main motivator was to free up staff to help customers and to promote stock and events. The key to moving to self-service is to build up staff skills and attitude in advance, to promote RFID to the public and to encourage staff to floor-walk. The customer panels were also immensely helpful in providing advice on the best ways to promote and introduce the new technology." In the self-service libraries the library counters have been removed so that the public are naturally drawn to the self-service machines; staff are available for queries via at least one information pod in each branch. The introduction of the machines also coincided with a makeover: libraries now have improved layouts and additional facilities such as self-service printing. The machines also handle money so that customers can pay for DVD loans or overdue fees themselves and Brent now has an average rate of 80% RFID usage across all its branches.

LIBRARY-PUBLISHER RELATIONSHIP IN CONTEXT OF RFID

1. While "closed" or "proprietary" library RFID systems may provide some privacy protection for library patrons today, this is not a viable long-term answer to the civil liberties risks. Closed systems rely fundamentally on "security through obscurity." Several forces seem poised to erode closed systems.
   — Book publishers will eventually use their own RFIDs. It likely won't happen for a few years, but the Book Industry Study Group is looking at RFIDs. To the extent that major retailers like Wal-Mart sell books and want suppliers to use RFIDs, the trend will gain momentum. An individual library would have a strong cost incentive to use whatever RFID system used by book publishers.

Many libraries already are parts of larger library systems or library consortia, and it should be expected that RFIDs will be interoperable within their own systems or consortia. Today's library RFIDs mainly operate in the high-frequency (HF) 13.56 MHz band, the most widely used of the RFID HF bands (UHF/868-915 MHz and microwave/2.45 GHz or 5.8 Ghz) because it's the global standard frequency for contactless smart cards or proximity cards. HF-band RFIDs have longer read ranges than low-frequency RFIDs, but don't need a power source. As a result, we should expect RFID readers/sensors for this frequency band to be very common, which will increase the likelihood that non-library RFID readers will be able to capture RFID data from library RFID tags.

a. Exploiting mass-production scale economies is a key major business goal of the RFID industry. It's unclear how long libraries will have market pull with RFID system vendors, because the industry needs to standardize and consolidate on a discrete set of options in order to sell as many RFID tags as possible for as many applications as possible globally. It's likely that the RFID industry and manufacturers, resellers, system integrators and government agencies will cooperate even more as the industry matures.

Libraries are one of the few government entities now using RFIDs, along with schools and the Defense Department. Unlike commercial RFIDs, government RFIDs (like library RFIDs) will be developed with persistence in mind: ID cards are an obvious example. Because the "kill" option won't be available- the privacy/civil liberties issues are more serious. Libraries need to think of themselves as part of the government and to think about the need for a deliberate policy process for government use of RFIDs, at least at the state and local level.

2. If libraries wish to influence RFID use and development toward privacy and security, the time is now. Libraries need to act collectively as soon as possible to protect their own and their patrons' interests, as well as to set an example for society.

Library-specific trends & developments

* Moving towards tag protocol and content standards,
* Moving away from vendors recommending as a primary solution for security purposes
* Vendors continue to explore ways to alternately secure media, while allowing RFID self-check; magnetic lockboxes around the case, cases with RFID-unlocking tabs, etc.
  - Continuing use of barcodes, despite early predictions that RFID would spell the end of barcode usage.
  - Continuing development of RFID products by integrated library systems.
  - Exploration of supply chain tagging, and use of

one tag through the life of the item.

RFID SOLUTIONS

For the concerns that library managers face each day, RFID technology brings solutions that magnetic stripe or barcode technology simply cannot deliver as effectively. The RFID tag is assigned and attached to the asset in similar ways, depending on the material, paper, plastics, etc. The advantages of RFID tags over barcode or magnetic stripe are as follows:

* Faster scanning of the data stored on the RFID tag,
* Simple and easy way for patrons to self-checking for material loans and returns,
* Superior detection rates,
* Significantly reduce the number of false alarms at the exit sensors (up to 75% less)
* High speed inventory - reducing time to by staff for 'shelf-reading' and other inventory activities,
* Automated return of materials that speeds up sorting of materials and re-shelving for the next patron to access and
* A longer lifecycle than a barcode.

EVALUATING RFID

Security feature : The same RFID tag used to manage inventory can also be used to protect it from theft. Current offerings provide the choice between a purely RFID solution, or RFID with an EM (electro-magnetic) add-on for theft.

Tag memory capacity : More memory is not necessarily better than less. It often correlates with price, and data transmission speed. It is important to consider what information library need to program into each tag, and then discuss with vendors.

Tag functionality:

* *Read/Write vs. Read Only*: Some vendors offer tags which can only be "written to" once. That is, once the tag is programmed, the information stored in the tag's memory cannot be changed. Alternatively, information stored in the memory of read/write tags can be updated as required.
* *Anti-collision*: All RFID vendors in the library market offer a product with anti-collision (the ability to read several tags simultaneously). However, the speed at which this can be performed, and the total number of tags that can be read, will vary. This relates specifically to inventory management with a hand-held reader, and check-in processes.

EAS (Electronic Article Surveillance) mechanism: As mentioned above, RFID can be used to prevent theft in the library. This approach varies from vendor to vendor - the security mechanism may be integrated into the chip itself, or security gates may be linked to a separate server which interrogates the database to conclude whether an alarm needs to be triggered.

Cost: It is expected to pay from US$0.85 to over US$1 per tag. The price of hardware (per unit) varies extensively from different suppliers. However, the infrastructure requirement also varies.

Standards: The emerging standard for library RFID solutions is to employ a frequency of 13.56MHz. However, no formal standards are currently in place.

Today's library RFIDs mainly operate in the high-frequency (HF) 13.56 MHz band and are very suitable for library application since they have longer read ranges, about 3 to 5 feet. Although RFID technology has matured and offers an ideal solution for libraries to replace their EAS and barcode systems for security and theft prevention, the picture is not all that smooth. There are several concerns and some are about patron privacy violations.

Large numbers of libraries have gone into RFID tagging for their inventory and have put in place best practices to inform their patrons about the use of technology and protection of their privacy. Though the unique advantages and flexibility of RFID is the good news, the technology is still not yet widely understood or installed in the library environment, and the cost/ROI models far from established. RFID, its application, standardization, and innovation are constantly changing. Its adoption is still relatively new and hence there are many features of the technology that are not well understood by the general populace. Developments in RFID technology continue to yield larger memory capacities, wider reading ranges, and faster processing. The interest in RFID as a solution to optimize further the automation and tracking of documents are gathering momentum at an increasing pace, with more libraries joining the trails. RFID is increasing in popularity among libraries, as the early adopters of this technology have shown that, it makes good economic sense, both for large and small libraries.

7
RFID and Barcodes

BEYOND BARCODES!

RFID systems originated in the 1940s, when the U.S. government used transponders to distinguish friendly aircraft from enemy aircraft. Through the 1970s, the federal government primarily used the systems for projects like tracking livestock and nuclear material. Radio tags have been used commercially for delivering packages, handling luggage, tracking food in supermarkets and monitoring highway tolls. Unlike bar codes, which need to be scanned manually and read individually, radio ID tags do not require line-of-sight for reading. Within the field of a wireless reading device, it is possible to automatically read hundreds of tags a second. Not only can these tags be read faster than bar codes, they also contain more information, so they can recall items more efficiently. The applications for this technology are seemingly endless. Radio ID tags can be installed in clothing labels, books, packaging, or even implanted beneath skin. MIT's Auto-ID Center was developing ways to use the tags in consumer product packages with blue chip companies such as Procter & Gamble, Wal-Mart, Gillette, Unilever, Target, Pepsi and Coca-Cola.

There are three types of RFID: high frequency (850-950 MHz and 2.4-5 GHz), intermediate frequency (10-15 MHz) and low frequency (100-500kHz). Low-frequency tags are used for applications such as security access and asset management, which require shorter read ranges. High-frequency systems are used for applications such as toll-collection and railroad car tracking, which require longer read ranges. While high-frequency tags transmit data faster and can be read from farther away, they also consume more power and are more expensive than low-frequency tags.

The Auto-ID Center was in the final stages of developing a standard way for RFID tags to communicate. The standard could be implemented in a mass production tag that would cost about 5 cents. Auto-ID researchers recommend placing a 64-bit (or 96-bit, depending on the version) number called an electronic product code (EPC) on an RFID tag. The EPC identifies every inventoried item with a unique serial number. During that, approximately 20 companies were participating in what may be the largest scale RFID test to date, prototyping tags based on the Auto-ID Center's protocol.

LOOKING BACK AT BAR CODES IN LIBRARY ENVIRONMENT

RFID is an identification technology; it does the same job as bar codes but offers potentially a lot more. It can therefore be fruitful to look back at bar code technology and see what we can learn from its application in library operations.

Barcode is a fundamental technology for library operations and flow management; it provides the coupling between the information system (the ILS, or Integrated Library System) and the physical flow of library material. In applications at the circulation desk, bar code technology has been proven to be robust, reliable, and efficient.

In the effort to extend bar code technology to self service stations, which is one major direction for achieving better efficiency in operations, the experiences have been less than satisfactory. The introduction of self service check-out stations was characterized by much work, very long delays in deliveries of functionality, and a long, long period of unreliable functionality. Trying to introduce self service check-ins was accompanied by even more difficulties.

Bar codes cannot be the signal feed for anti-theft systems so the usual combination is to have a combination of bar codes for identification, and some electromagnetic (EM) based anti-theft system. In self service stations the identification system must interact with the anti-theft system so that items have the correct EM state after the circulation transactions, i.e. checkedout item must be deactivated, and returned items activated. As for the signal source for the anti-theft system, there is a danger of a technology (or system) trap: once the magnetic device is attached to the library item it should, ideally, not be possible to take it away. This, however, makes it difficult to switch to another EM security system, if the signal source cannot be used in the alternative solution. It is also possible that the option of "ignoring" old magnetic devices (and just add new ones) is not realizable since remaining old devices can interfere with the new ones.

As the RFID refers to a set of small electronic devices that consist of a small chip and an antenna, in general, the small chip carries 2,000 bytes of data or less. This RFID device serves the same purpose as a bar code or a magnetic strip on the back of a credit card or ATM card by providing a unique identifier for that object. The RFID device must be scanned to retrieve the identifying information as a bar code or magnetic strip must be scanned to get the information. RFID technology is similar to the bar code identification systems; the difference between RFID and bar code technology is that RFID does not rely on the line-of-sight reading that bar code scanning requires to work.

HOW RFID WORKING MAKES DIFFERENCE WITH BARCODE!

The electromagnetic or electrostatic coupling in the RF (radio frequency) portion of the electromagnetic spectrum is used to transmit signals in RFID. The system consists of an antenna and a transceiver to read the radio frequency and transfer the information to a processing device (reader) and a transponder (RF tag). This RF tag contains the RF circuitry and information which is to be transmitted. The antenna supports the integrated circuit to transmit the information to the reader. The reader converts the radio waves reflected back from the RFID tag into digital information. This information is passed on to computers to analyze the data.

The tags that hold the data are of two different types i.e. Passive and Active tags. The Passive tags use the radio frequency from the reader to transmit their signal. Passive tags generally have their data permanently burned into the tag when it is made, although some can be rewritten. Active tags are more sophisticated. They have on-board battery for power to transmit their data signal over a greater distance and power random access memory (RAM) giving them the ability to store up to 32,000 bytes of data. Unlike a bar code label which requires line of sight reading and can only have a single code for all items of a particular product, an RFID tag can identify each individual unit of that product even when it is on the move, besides allowing the data on it to be updated

1. Limitations of Barcodes

It is felt that RFID is so expensive. The use of bar code is something people have become very familiar with, and it is difficult to find an application that has not benefited from the use of bar code. However, bar codes do have some very important limitations that RFID can overcome. Using the two hand in hand can give some significant benefits to a system. The second question is a difficult one to answer, but in many questions the answer is "how can you afford to NOT implement RFID".

There are two areas that bar codes do not perform well at : changing data and harsh environments. A bar code does not stand up to an automobile paint show very well. And once a bar code is printed, it will persist. Any organization can print a new one, but it cannot change the first one.

RFID tags have the capability to have the data changed on them many times depending on the specific type. It can store information and update it as a tag moves through a process, keeping the important information with the tag (and the item) and so making it available at any point in its life. Now some items do not need information changed frequently and this may well be an example of when a bar code will do the job needed. Now think about the item being used in a harsh environment. The same tag will work in most situations, hot, cold, dirty etc. where a standard bar code label might have problems (direct thermal printed labels don't like the heat too much) and so give the flexibility needed to incorporate into a total system.

2. In-depth understanding of RFID Over Barcodes

RFID tags and barcodes both carry information about products. However, there are important aspects related with these two technologies:

* Barcode readers require a direct line of sight to the printed barcode; RFID readers do not require a direct line of sight to either active RFID tags or passive RFID tags.
* RFID tags can be read at much greater distances; an RFID reader can pull information from a tag at distances up to 300 feet. The range to read a

barcode is much less, typically no more than fifteen feet.

* RFID readers can interrogate, or read, RFID tags much faster; read rates of forty or more tags per second are possible. Reading barcodes is much more time-consuming; due to the fact that a direct line of sight is required, if the items are not properly oriented to the reader it may take seconds to read an individual tag. Barcode readers usually take a half-second or more to successfully complete a read.
* Line of sight requirements also limit the ruggedness of barcodes as well as the reusability of barcodes. (Since line of sight is required for barcodes, the printed barcode must be exposed on the outside of the product, where it is subject to greater wear and tear.) RFID tags are typically more rugged, since the electronic components are better protected in a plastic cover. RFID tags can also be implanted within the product itself, guaranteeing greater ruggedness and reusability.
* Barcodes have no read/write capability; that is, one cannot add to the information written on a printed barcode. RFID tags, however, can be read/write devices; the RFID reader can communicate with the tag, and alter as much of the information as the tag design will allow.
* RFID tags are typically more expensive than barcodes, in some cases, much more so.

3. Advantages of RFID over Barcodes

A significant advantage of RFID devices over the others is that the RFID device does not need to be positioned precisely relative to the scanner. It is necessary to make sure that a barcode can be read. The credit cards and ATM cards must be swiped through a special reader. For comfort the RFID devices works within a few feet (up to 20 feet for high-frequency devices) of the scanner. It would be able to query all of the RFID devices and read the information immediately.

RFID technology has been available for more than fifty years. It has only been recently that the ability to manufacture the RFID devices has fallen to the point where they can be used as a "throwaway" inventory or control device. For example, Alien Technologies recently sold 500 million RFID tags to Gillette at a cost of about ten cents per tag. One reason that it has taken so long for RFID to come into common use is the lack of standards in the industry. Most companies invested in RFID technology only use the tags to track items within their control; many of the benefits of RFID come when items are tracked from company to company or from country to country.

Though RFID and barcodes have similarities as they are both data collection technologies yet they differ significantly in many areas. This comparison primarily focuses on the advantages of RFID over barcodes; RFID will probably not completely replace barcode technology. Barcodes offer some advantages over RFID, most notably the low cost.

Table : RFID vs. Barcodes Comparison

	RFID	Barcode
Line of Site	Not required (in most cases)	Required
Read Range	Passive RFID: - Up to 40 feet (fixed readers) - Up to 20 feet (handheld readers) Active RFID: - Up to 100's of feet or more	Several inches up to several feet
Read Rate	10's, 100's or 1000's simultaneously	Only one at a time
Identification	Can uniquely identify each item/asset tagged	Can typically only identify the type of item (UPC Code) but not uniquely

Read/Write	Many RFID tags are Read/Write	Read only
Technology	RF (Radio Frequency)	Optical (Laser)
Interference	Like the TSA (Transportation Security Administration), some RFID frequencies don't like Metal and Liquids. They can cause interfere with certain RF Frequencies.	Obstructed barcodes cannot be read (dirt covering barcode, torn barcode, etc.)
Automation	Most "fixed" readers don't require human involvement to collect data (automated)	Most barcode scanners require a human to operate (labor intensive)

Source: http://www.inlogic.com/rfid/rfid_overview.aspx

Comparing the use of both technologies for books management

	Reader Management	Book Management
Bar-code	Reader number (bar-code) printed on reader's card	Bar-code sticked to outside of book
Pros / Cons	+ Reading reliable - Easy to duplicate	+ Reading reliable - Theft control inefficient - Inventory difficult and cumbersome - Non-modifiable - Single item check-out - Line-of-Sight Technology
RFID chip	Reader number written inside reader's RFID card	RFID tag sticked inside book
Pros / Cons	+ Very difficult to duplicate - May involve reading difficult to understand for nonspecialists	+ Theft control (offline gates) + Easier inventory process problems, (Handheld device)

	+ Fewer resources for check-in/out (Self-Check Station) + Contains more information (name of library, type of book) - Book tagging process cumbersome - Bar-codes are kept for compatibility with non-RFID libraries

(***Source:*** RFID - Technology and Applications in Libraries, by Nils Rinaldi IFLA-PAC Conference, 26th March 2008)

COMMON PROBLEMS WITH RFID

Some common problems with RFID are reader collision and tag collision. Reader collision occurs when the signals from two or more readers overlap. The tag is unable to respond to simultaneous queries. Systems must be carefully set up to avoid this problem. Tag collision occurs when many tags are present in a small area; but since the read time is very fast, it is easier for vendors to develop systems that ensure that tags respond one at a time.

1. When doesn't RFID make sense

There are obviously situations where it doesn't make sense, either practically or economically, to use RFID. Tagging low value items with a high cost single use tag is probably not a good idea. Using an RFID system in an application with large amounts of ferrous metal may pose a challenge. But in those situations where the tag is re-used the answer is almost always a big YES. Asset tracking, reusable containers, harsh environments, and non line-of-sight applications for example are classic candidates for an RFID system and user will find that initial cost issues will disappear when he/she looks at the return on investment.

2. Summing up the comparison between Barcode and RFID

* Information can be read from RFID tags much faster than from barcodes
* Several items in a stack/counter can be read at the same time using RFID
* Items do not have to be handled one-by-one nor removed from the shelves
* Inventory-taking is no longer a tedious operation
* RFID can stand more than 10,000 read/write
* RFID can have theft bit which can be in two states "ON/OFF"
* Shelf verification/rectification can be done on daily basis
* More information can be written in the RFID tag on incremental basis
* Need not open/remove books to capture information
* Items are identified on upper and lower shelves more comfortably

THE FUTURE OF RFID

RFID is said by many in the industry to be the frontrunner technology for automatic identification and data collection. The biggest, as of yet unproven, benefit would ultimately be in the consumer goods supply chain where an RFID tag attached to a consumer product could be tracked from manufacturing to the retail store right to the consumer's home. Many see RFID as a technology in its infancy with an untapped potential. While we may talk of its existence and the amazing ways in which this technology can be put to use, until there are more standards set within the industry and the cost of RFID technology comes down we won't see RFID systems reaching near their full potential anytime soon.

The success of RFID in the mass market ultimately depends upon tying the technology to an open standard network like the Internet. The Internet is absolutely crucial for all this. The Internet enables to put all the data on the network, not the tag, and that makes tags cheaper, better, faster. Organizations can make the tags cheap by transmitting most of the information on the Internet. Eventually the technology could be used to improve health care, public safety and recycling. This technology can help save money properly, lives and the environment. Though this technology seemed, yet cost and lack of standards have been a major barrier for widespread adoption. Once costs come down, RFID could become as pervasive as bar codes are today.

8

RFID Privacy Issues

PRIVACY! OR PERVASIVE!

Before RFID can be understood completely, it is essential to understand how Radio Frequency communication occurs. RF (Radio Frequency) communication occurs by the transference of data over electromagnetic waves. By generating a specific electromagnetic wave at the source, its effect can be noticed at the receiver far from the source, which then identifies it and thus the information. The RFID tag which contains the tagged data of the object generates a signal containing the respective information which is read by the RFID reader, which then may pass this information to a processor for processing the obtained information for that particular application

Thus, an RFID System can be visualized as the sum of the following three components:

* RFID tag or transponder
* RFID reader or transceiver
* Data processing subsystem

An RFID tag is composed of an antenna, a wireless transducer and an encapsulating material. These tags can be either active or passive. While the active tags have on-chip power, passive tags use the power induced by the magnetic field of the RFID reader. Thus passive tags are cheaper but with lower range (<10mts) and more sensitive to regulatory and environmental constraints, as compared to active tags. An RFID reader consists of an antenna, transceiver and decoder, which sends periodic signals to inquire about any tag in vicinity. On receiving any signal from a tag it passes on that information to the data processor. The data processing subsystem provides the means of processing and storing the data.

RFID systems can also be differentiated based on the frequency range it uses. The common ranges are Low-Frequency (LF: 125 - 134.2 kHz and 140 - 148.5 kHz), High-Frequency (HF: 13.56 MHz) and Ultra-High-Frequency (UHF: 868 MHz - 928 MHz). Low-frequency systems have short reading ranges and lower system costs. They are most commonly used in security access, asset tracking, and animal identification applications. High-frequency systems, offering long read ranges (greater than 90 feet) and high reading speeds, are used for such applications as railroad car tracking and automated toll collection. However, the higher performance of high-frequency RFID systems incurs higher system costs.

RFID TECHNOLOGY AND PRIVACY

The use of RFID technology has engendered considerable controversy and even product boycotts by consumer privacy advocates. The two main privacy concerns regarding RFID are:

* Since the owner of an item will not necessarily be aware of the presence of an RFID tag and the tag can be read at a distance without the knowledge of the individual. This situation can lead to gather sensitive data without consent.
* If a tagged item is paid for by credit card or in

conjunction with use of a loyalty card, then it would be possible to indirectly deduce the identity of the purchaser by reading the globally unique ID of that item (contained in the RFID tag). This is only true if the person doing the watching also had access to the loyalty card data and the credit card data, and the person with the equipment knows where the owner is going to be.

The concerns raised by the above may be addressed in part by use of the Clipped Tag. The Clipped Tag is an RFID tag designed to increase consumer privacy. The Clipped Tag has been suggested by IBM researchers Paul Moskowitz and Guenter Karjoth. After the point of sale, a consumer may tear off a portion of the tag. This allows the transformation of a long-range tag into a proximity tag that still may be read, but only at short range - less than a few inches or centimeters. The modification of the tag may be confirmed visually. The tag may still be used later for returns, recalls, or recycling.

However, read range is both a function of the reader and the tag itself. Improvements in technology may increase read ranges for tags. Having readers very close to the tags makes short range tags readable. Generally, the read range of a tag is limited to the distance from the reader over which the tag can draw enough energy from the reader field to power the tag. Tags may be read at longer ranges than they are designed for by increasing reader power. The limit on read distance then becomes the signal-to-noise ratio of the signal reflected from the tag back to the reader.

IMPACT OF RFID ON CONSUMER PRIVACY

This tiny wireless technology has the potential to radically transform the commerce world. It consists of an inexpensive chip, often smaller than a grain of sand, which can be read up to several meters away. The hope among retailers is that the technology can be used as a next-generation barcode, automating inventory levels, and thus cutting costs for manufacturers and retailers. While barcode-type RFID tags are not likely to reach consumers on a regular basis in the near future, there is little doubt that with the success of early trials, they eventually will. Consequently, concerns about privacy infringement have vital practical applications.

1. Few developments

In January 2004 privacy advocates from CASPIAN and the German privacy group FoeBuD were invited to the METRO Future Store in Germany, where an RFID pilot project was implemented. It was uncovered by accident that METRO "Payback" customer loyalty cards contained RFID tags with customer IDs, a fact that was disclosed neither to customers receiving the cards, nor to this group of privacy advocates. This happened despite assurances by METRO that no customer identification data was tracked and all RFID usage was clearly disclosed. In 2004-2005, the Federal Trade Commission Staff conducted a workshop and review of RFID privacy concerns and issued a report recommending best practices. RFID was one of the main topics of 2006 Chaos Communication Congress (organized by the Chaos Computer Club in Berlin) and triggered a big press debate. Talks showed how the first real world mass application of RFID technology at the 2006 FIFA Soccer World Cup worked.

RFID already has the capability to determine the distance of a tag from the reader location. With such technology already available, it is not difficult to imagine a situation in which retailers could determine the location of individuals within their store, and thus target specific advertisements to that customer based upon past purchases. In effect, that store would be creating a personal log of buyer's past purchases, his/her shopping patterns, and ultimately his/her behavioral patters. While such information gathering would be considered intrusive enough by many consumer's standards, the danger that such information could be sold to other retailers, could create potentially devastating information vulnerabilities. While some RFID critics have pointed out that the technology could lead to some sort of corporate "Big Brother," there is a more widespread concern that allowing RFID to develop without legal restrictions will eliminate the possibility for consumers to refuse to give such information to retailers.

2. Specific Privacy Concerns

RFID tags differ from conventional barcode tags in a number of ways. It is these differences that create the benefit of adopting the technology, while simultaneously creating the greatest concern over the privacy issues involved. For example, under today's barcode technology, a pack of item sold in a particular place has the same barcode as a pack sold in some other place. With RFID, however, each pack would have a unique ID code which could be tied to the purchaser of that item when they use an "item registration system" such as a frequent shopper card or a credit card.

Continuing with the above example, the purchaser could then be tracked if he/she ever entered that same store again or perhaps more frightening, if they entered any other store with RFID reading capability. Because, unlike a barcode, RFID tags can be read from much greater distances and the reading of such devices is non-directional. This means that if buyer enters a store with a pack of item in his/her pocket, the reader can identify that pack of particular item, the time and date on which it is bought, where buyer bought it, and how frequently a particular buyer come into the store. If buyer used a credit card or a frequent shopper card to purchase it, the manufacturer and store could also tie that information to buyer's name, address, and e-mail.

As the technology behind RFID advances, the potential for privacy infringement does as well. Privacy is the biggest single research area. It is needed to make sure the technology is not a threat to anyone's privacy and make sure everyone is comfortable with that. An example can be found at the Auto-ID center which has designed privacy protection measures into the EPC system. The chips contain no useful data — just a unique code number that refers to information held remotely on the Internet. Access to this information is restricted and controlled. The tags cannot be read through walls or from more than about five feet away. The Uniform Code Council, creator of the bar code, has endorsed the Auto-ID Center's efforts.

3. Privacy Problems

However, privacy concerns associated with item-level tagging is another significant impediment to library use of RFID tags. The problem with today's library RFID systems is that the tags contain static information that can be relatively easily read by unauthorized tag readers. This allows for privacy issues described as "tracking" and "hotlisting." Tracking refers to the ability track the movement of a book (or person carrying the book) by "correlating multiple observations of the book's bar code" or RFID tag. Hotlisting refers to process of building a database of books and their associated tag numbers (the hotlist) and then using an unauthorized reader to determine who is checking out items on the hotlist. Current standards (ISO 15693) apply to container-level tagging used in supply chain applications, and do not address problems of tracking and hotlisting. Next generation tags (ISO 18000) are designed for item-level tagging. The newer tags are capable of resolving many of privacy problems of today's tags. However, no library RFID products are currently available using the new standard.

4. Counter-Measures

Despite these and other such dangers, however,

there are some steps being taken to mitigate these privacy issues. For example, a recent proposal would require that all RFID-tagged products be clearly labeled. This would give consumers the choice to select products without RFID, or at a minimum to recognize that the items they select are being tracked. For those unsatisfied with disclosure, there also exist a growing number of products designed to limit their exposure to RFID tagged products. One such product is "Kill Codes," a command which turns off all RFID tags immediately as the consumer comes into contact with them, thus entirely eliminating the effectiveness of the technology. Another countermeasure against RFID privacy invasion is "RSA Blocker Tags," which try to address privacy concerns while maintaining the integrity of the product. Under this technology, the item can only be tracked by that store's authorized reader, meaning that customers cannot be tracked outside of the store in which they purchased the item.

PRIVACY PROTECTIONS FOR RFID

The Privacy Act of 1974 articulates certain principles for protecting privacy. Since its passage, these principles have been applied to specific technologies and practices. Bruening (2004) describes the underlying principles as follows:

Notice: Information collection and use should be open and transparent.

Purpose specification: Personal data should be relevant to the purposes for which it is collected.

Use limitation: Data should be used only for the purpose for which it was collected

Accuracy: Personal data should be accurate, complete, and timely.

Security: Personal data should be protected by reasonable security safeguards against risk of loss, unauthorized access, destruction, use, modification or disclosure.

Access: Individuals should have a right to view all information that is collected about them to correct data that is not timely, accurate, relevant or complete.

Accountability: Record keepers should be accountable for complying with fair information practices.

Privacy concerns are the main drawback to the technology. The very usefulness of the system is what makes it vulnerable to privacy abuses. There are two types of tags on the market:

1. **Read-only capability type tag** : The information stored on the tag is only that which was entered at the time the item was processed. These tags have limited privacy issues related to them as no patron information is added to the item's RFID tag.

2. **Read/write type of tag** : These tags can be both read by the RFID readers and have information added to the tag by them, such as the patron's name and other information. The potential for a privacy breach by such a tag is possible. However, all systems are structured differently; encoding has differences required by the individual institutions. A compliant reader could read the information off the tag but the information would be in an encoded form. The theft would have to be by someone who understood how that particular organization encoded its material.

The American Library Association policy, adopted in January 2005, calls for careful development of privacy policies prior to the planning of a RFID implementation. The Ontario Information and Privacy Commissioner wrote a report on Guidelines for Using RFID Tags in Ontario Public Libraries. This report outlined some of the concerns and policy issues each library needs to develop prior to starting an RFID implementation. The report also included a best practices outline.

PRIVACY AND SECURITY

RFID tags have the potential to revolutionize the shopping experience by bringing customers' targeted products and allowing retailers and manufacturers to track purchases and customers' behavior more accurately and cost-effectively. The concern, however, is that if people are not aware and careful about the potential abuses of such technologies early on, they may fail to incorporate them at a time when the laws and mores of such a system are still developing, ultimately suffering the consequences later on.

1. Few approaches towards RFID Security and Privacy

Security and privacy is a major issue for many elements of technology these days. RFID security and RFID privacy are no exception. RFID security and RFID privacy are major features of the technology and they feature highly in the way in which RFID systems are designed, set up and operated. RFID security is an issue from several viewpoints including prevention of the system being hacked, to issues of personal security or privacy and access to personal location information, etc. Both RFID security and RFID privacy go hand in had, issues of one affecting the other.

A major concern over the increase in use of RFID tags is the personal security associated with the illicit tracking of RFID tags. Private organisations have expressed their concern over this with the increasing use of electronic product code tags being embedded in consumer products. There are a number of ways in which these RFID security issues arise:

* The unique identifiers within RFID tags can be used for profiling and identifying consumer and Individual patterns.
* Stealth readers can track people with RFID tags on them.

* Hidden tags could be placed within on or within an item to enable stealth tracking to be undertaken.

To help overcome these issues a number of approaches can be adopted:

Blocker tags: These tags spam any unauthorized users into assuming that there are many tags in the vicinity and thereby preventing access to any tags that may be on the individual.

Kill switches: When consumer items are purchased, or if an RFID tag needs to be deactivated for any reason, tags can be disabled. This function is incorporated into many newer RFID tags, although not all. It also does not stop the possibility of illicit tags being placed and used for tracking.

While these measures are available, the possibility of RFID security for individuals is something which can derive some considerations.

2. RFID tag cloning

There are many other issues associated with RFID security, one of which is RFID tag cloning. RFID security systems need to be able to prevent cloning as this would open the overall system to a variety of forms of security attack. Typically when RFID tag cloning occurs, the responses of RFID tags are received by rogue monitors. Information received can then be used to replicate tags. To enable RFID security to overcome this vulnerability, cryptographic techniques are used and embedded into the chips. A number of approaches may be adopted:

Rolling code approach: This approach to RFID security uses a scheme where the identifier given by the RFID tag changes after each read action. It requires the RFID reader and RFID tag to have the same algorithm for changing the identifier. If multiple readers are used, they must be linked so that tracking can occur.

Challenge response authentications: These

systems use cryptographic principles. Here the reader issues an enquiry to the tag which results in a response, but as secret tag information is never sent over the interface between the RFID reader and tag the system cannot be compromised. Both reader and tag compute information from internal cryptographic algorithms, and the results are sent. The correct responses required for a successful information interchange. The system is essentially the same as encrypting data to send over a normal radio link.

In view of the additional processing required, the tags have a very much higher cost. Deployment of these RFID secure tags is limited to areas where the cost can be justified. RFID security and RFID privacy both remain as issues. In many cases the limited range provides the level of security required by many. Also there is often not a direct gain that can be made by criminals, so RFID security is not an issue in the same way as that for credit cards.

9
RFID Standards

THRESHOLD!

Internet information services and consultants, RFID seems to be "the new technology hype" which "will revolutionize business performance across supply chains". The supporters argue that RFID significantly reduces costs, increases the transparency and hence improve the visibility of the entire supply chain, leading one step further towards the achievement of the truly integrated and virtual supply chain. In contrast, the critics draw the attention on the huge technical challenges such as integration with the existing IT infrastructure and the even more substantial organizational changes required by the adoption of RFID. Major obstacle for the widespread adoption of RFID is RFID standards. In a global business environment, the lack of interoperability between systems based on RFID technology in different parts of the world deter users to make large investments in a technology that has to be used on a global basis.

The exploitation of technology in the form of products and services, national and international usage for domestic, invariably requires the support of standards. They facilitate compatibility and interoperability within the applications arena. The need generally arises for a standard or standards to support the 'diffusion process' both in terms of application growth and expansion of the supplier base for supporting technology. In such circumstances the proprietary specification may be seen as the vehicle for a de facto standard. Proposals for standards may also evolve in other ways, be developed and accepted through a formalised standardisation process, as industry or national standards. There are four main types of standards relevant to Total Asset Visibility:

Technology Standards: dealing with the technological features, such as the air interface communications format and data exchange protocols that have to be agreed to assure compatibility or interoperability in systems produced by different manufacturers or system providers.

Application Standards: dealing with agreements on the way or ways in which technology or systems are to be used in particular applications to ensure consistent usage in a specified manner. Many devices only require technology standards. Application standards often set parameters and associated values, which constrain the associated technology standard in terms of performance or other features. It may incorporate the data standard, or refer to an external data standard.

Data Standards: dealing with the agreements on the way data is structured for compatibility and interoperability requirements.

Conformance Standards: dealing with agreements that specify the way in which systems are to perform to be acceptable with respect to particular performance or operational criteria.

Due to the 'radio' or electromagnetic nature of RFID, the conformance requirements extend to particular regulatory demands with respect to electromagnetic spectrum usage. RFID data carriers (tags or transponders) and associated systems (may be viewed as short range devices; SRDs), share spectrum with other SRDs such as telemetry systems and radio alarms. They are also required to comply with following requirements concerning:

* licensed or unlicensed operational parameters,
* electromagnetic compatibility,
* interference immunity and
* safety.

These regulatory aspects can differ between countries and regulatory authorities. Regulations relating to short range devices can generally be distinguished for particular bands and power levels and are invariably specified at national level according to nationally defined needs. As a consequence national regulatory specifications can often be seen as a source of non-harmonisation when viewing the needs for international standards.

It's commonly said that there are no standards in RFID. In fact, there are many well-established standards and a few emerging standards. Standards are critical for many RFID applications. A great deal of work has been going on to develop standards for different RFID frequencies and applications. The proposed RFID standards deal with:

* air interface protocol (the way tags and readers communicate),
* data content (the way data is organized or formatted),
* conformance (ways to test that products meet the standard) and
* applications (for example, how standards are used on shipping labels).

The International Organization for Standardization (ISO) has created standards for tracking cattle with RFID.

ISO 11784 defines how data is structured on the tag. ISO 11785 defines the air interface protocol. ISO has created a standard for the air interface protocol for RFID tags used in payment systems and contactless smart cards (ISO 14443) and in vicinity cards (ISO 15693). It also has established standards for testing the conformance of RFID tags and readers to a standard (ISO 18047), and for testing the performance of RFID tags and readers (ISO 18046). Using RFID to track goods in open supply chains is relatively new and fewer standards have been finalized. ISO has proposed standards for tracking 40-foot shipping containers, pallets, transport units, cases and unique items. These are at various stages in the approval process.

The standard situation was complicated by the fact that the Auto-ID Center, which developed Electronic Product Code technologies, chose to create its own air interface protocol for tracking goods through the international supply chain. The Auto-ID Center also wanted its RFID system to be global and to be based on open standards. It needed to be global because the aim was to use it to track goods as they flowed from a manufacturer in one country or region to companies in other regions and eventually to store shelves. For Company A to read a tag put on a product by Company B, the tag had to use a standardized air interface protocol. The Auto-ID Center developed its own protocol and licensed it to EPCglobal on the condition that it would be made available royalty-free to manufacturers and end users. The center was charged with developing a network architecture also — a layer integrated with the Internet— that would enable anyone to look up information associated with a serial number stored on a tag. The network, too, needed to be based on open standards used on the Internet, so companies could share information easily and at low cost. The Auto-ID Center developed its own UHF protocol. Originally, the center planned to have one protocol that could be used to communicate with different classes of tags.

ISSUES

The issues surrounding RFID technologies can be categorized as following :

1. The RFID market has massive amount of diverse players such as chip makers, transponder manufacturers, system integrators or consultancies, all of whom offer different, and generally proprietary, products and services. Available systems consist of different frequency ranges, transfer modes, etc. For a potential customer, it is difficult to acknowledge the distinct benefits and disadvantages of these different RFID solutions.
2. A selection of different RFID systems has to be done by the users depending on the organisational specific process and technological requirements.
3. Due to the fragmented market, the total cost of RFID implementation is not transparent. A RFID implementation cost analysis has to take into consideration not only the investment in the transponders and readers, but also other cost drivers, such as peripheral systems, software and integration efforts.
4. RFID technologies require a huge effort in terms of standardization. RFID standards are a major issue in securing the high investments in RFID technology on different levels (e.g. interface protocol, data structure, etc.).

An RFID system can utilize a few standards. The problem has been that there is no one universally accepted standard. Competing standards have been one of the more difficult issues for RFID, and as a result, most RFID applications have been closed systems. Standards and specifications may be set at the international, national, industry or trade association level, and individual organizations may term their own specifications as "standard." Many industry standards and specifications set by individual organizations are based on international standards to make implementation and support easier and to provide a wider choice of available products. The two largest drivers for RFID have been Wal-Mart and the Department of Defense (DOD). Both have issued mandates for their top suppliers to use RFID technology when shipping products to their distribution centers. They are both looking to accomplish the same thing, but have a slightly different long-term outlook. Like many other technologies, RFID standards are used. RFID standards, as any other standards enable manufacturers to make the same products for a variety of markets and in this way gain the economies of scale. RFID standards also enable products from different manufacturers to operate together. One example may be for tags which are used in very large quantities and manufacturers may want to source from two suppliers to provide reliability.

RFID STANDARDS BODIES

There are two main international RFID standards bodies or standardisation bodies:

* ISO - International Standards Organisation
* EPCglobal - Electronics Product Code Global Incorporated

Although these two organisations provide the main RFID standards organisations, there is also a plethora of other standards that apply to niche areas of RFID.

In terms of the standardisation organisations, ISO is the longest established. In 1996, it set up a joint committee with IEC to look at standardisation for RFID technology. The ISO RFID standards fall into a number of categories according to the aspect of RFID that they are addressing. These include: air interface and associated protocols; data content and the formatting; conformance testing; applications; and various other smaller areas. In addition to the ISO RFID standards, there are also the standards from EPC Global.

AUTO-ID TAG STANDARDS

In order to be able to standardise the RFID tags, the Auto-ID Center devised a series of classes for RFID tags. was generated and these still form the basis for a developed system of RFID tag classes seen today:

Class 0: Basic read-only passive tag using backscatter where the tag was programmed at the time the tag chip was made.

Class 1: Basic read-only passive tag using backscatter with one-time non-volatile programme capability.

Class 2: Passive backscatter tag with up to 65k of read-write memory.

Class 3: Semi-passive tag with up to 65 k read-write memory and a battery incorporated to provide increased range.

Class 4 : Active tag using a with battery to enable extra functionality within the tag and also to provide power for the transmitter.

Class 5: An active tag that provides additional circuitry to communicate with other class 5 tags.

The responsibilities for Class 0 and Class 1 RFID tag definitions and standards was handed on to EPC Global in 2003. Although other newer RFID tag standards and now available, reference is still made to these original tag classes.

GEN2 RFID TAG STANDARD

In 2004, EPCglobal began the creation of a second generation protocol, often referred to as EPCglobal Gen2. While the EPCglobal Gen2 RFID standard is not backwards compatible with the Class 0 and Class 1 tags, it aims to provide a worldwide RFID tag standard that is compatible with ISO standards. The ISO and the EPC Global have both been leading figures in this debate. The ISO has their 18000 standard and the EPC Global Center has introduced the EPC standard. Wal-Mart has decided to use the EPC standard, where the DOD wants to use the EPC for general purposes, but use the ISO standard for air interface.

EPC standard for air interface is not compatible with the ISO 18000 UHF (Part 6) standard. Both, the EPC and ISO 18000 (Part6) standards, deal with the tracking of merchandise through the supply chain. The ISO 18000 (Part 6) standard only deals with air interface protocols, whereas the EPC standard also includes data structure. The desire is for these two protocols not to be mutually exclusive. There are several evolutions to the EPC standard. Class 1-Generation 1 is the current version of EPC. It is not backward compatible with Class 0. Generation 2 was hoped to be backward compatible with Class 0 but merging with the ISO 18000 standard will be difficult, if not impossible. The EPC standard was developed for carton and pallet tracking within the supply chain.

RFID STANDARDS CREATION

There are two competing initiatives in the RFID standardization arena: ISO and EPC Global. Additionally, there are also a number of special interest groups including industry specific such as the American Trucking Association in the transport industry, the NFC forum in consumer electronics, mobile devices and computer industry or the Automotive Industry Action Group in the automotive industry that seek to influence RFID standards development.

THE ISO APPROACH

RFID standards first come to scene during the early 1990s, when the (newly created) CEN TC225 committee on bar coding focused the attention on automatic ID techniques in general. During the early 1990s, the standardisation activity on automatic ID techniques was mainly carried out in Europe within the CEN standard body (TC225 committee), with little involvement from the US. However, during the 1995, a joint ISO IEC JTC1 committee - the SC31 - was set up for standardisation of automatic identification techniques generally drawing from the earlier work on RFID standards within CEN. Another influence on the RFID work within ISO was the work on the GTag initiative for RFID standardisation of asset tracking and logistics which was launched by UCC and EAN in 2000 along with input from international companies including Philips Semiconductors, Intermec, and Gemplus.

The members of the SC31 committees were the representatives of the national standard bodies such as in UK the BSI IST34 committee on far coding, including the same people who tend to participate in CEN TC225. Three different levels of representativeness (and thus interests) can be identified in the ISO process: the individual, the organisational, and the national level. RFID ISO standards cover 4 different areas: technology (e.g. ISO 18000 series), data content (e.g. ISO 15418), conformance and performance (e.g. ISO 18046), and application standards (e.g. ISO 10374), and the focus here is on the technology standards, the ISO 18000 series, and in particular the air interface standards, which are developed within the SC31 committee. ISO standards are generic, being able to be supported by any system and in any context, irrespective of the data that is being carried.

ISO Standards for Proximity Cards

ISO 14443 for *"proximity"* cards and ISO 15693 for *"vicinity"* cards both recommend 13.56 MHz as its carrier frequency. These standards feature a thinner card, higher memory space availability and allow numerous cards in the field to be read almost simultaneously using anti-collision, bit masking and time slot protocols.

* **ISO 14443 proximity cards** offer a maximum range of only a few inches. It is primarily utilized for financial transactions such as automatic fare collection, bankcard activity and high security applications. These applications prefer a very limited range for security.
* **ISO 15693 vicinity cards, or Smart Tags,** offer a maximum usable range of out to 28 inches from a single antenna or as much as 4 feet using multiple antenna elements and a high performance reader system.

ISO Standards for RFID Air interface

The ISO 18000 series is a set of proposed RFID specifications for item management that could be ratified as standards during 2004. The series includes different specifications that cover all popular frequencies, including 135KHz, 13.56 MHz, 860-930 MHz and 2.45 GHz.

* **18000 – 1 Part 1:** Generic Parameters for Air Interface Communication for Globally Accepted Frequencies
* **18000 - Part 2**: Parameters for Air Interface Communications below 135 KHz o ISO standard for Low Frequency
* **18000 - Part 3**: Parameters for Air Interface Communications at 13.56 MHz o ISO standard for High Frequency o Read \ Write capability
* **18000 - Part 4**: Parameters for Air Interface Communications at 2.45 GHz o ISO standard for Microwave Frequency
  — Read \ Write capability

- **18000 - Part 5:** Parameters for Air Interface Communications at 5.8 GHz
- **18000 - Part 6:** Parameters for Air Interface Communications at 860 – 930 MHz
  - ISO standard for UHF Frequency
  - Read \ Write capability
  - Targeted for same markets as EPC standards.
- **18000 – Part 7:** Parameters for Air Interface Communications at 433.92 MHz o Manifest tag for Department of Defense (DOD)

Note: In the UHF arena, the EPC standard for air interface is not compatible with the ISO 18000 UHF (Part 6) standard. The ISO 18000 (Part 6) standard only deals with air interface protocols, whereas the EPC standard also includes data structure. The desire is for these two protocols not to be mutually exclusive.

ISO Standards for Animal Identification

- **ISO 11748 / 11785:** Standard for Animal Identification

ISO Supply Chain Standards

These are used to identify different types of logistics containers and packaging, in addition to individual items.

- **ISO 17358** - Application Requirements, including
- Hierarchical Data Mapping
- **ISO 17363** - Freight Containers
- **ISO 17364** - Returnable Transport Items
- **ISO 17365** - Transport Units
- **ISO 17366** - Product Packaging
- **ISO 17367** - Product Tagging (DoD)
- **ISO 10374.2** - RFID Freight Container Identification

THE EPC GLOBAL APPROACH

The need for the creation of a formal standard body was recognized that would take over the standardisation and commercialisation work within Auto-ID. A new entity was created in October 2003, the EPC Global as a joint venture between UCC and EAN. Whereas Auto-ID would continue to research RFID technologies, EPC Global focuses on standardisation activities, as well as their commercialization.

In contrast with ISO RFID standards which are generic standards, EPC standards are specific. EPC standards describe the tag and the air interface depending on the data being carried. EPC standards prescribe the physical implementation of the tags and readers, rather then specifying their generic characteristics. The standards are also much more limited in their scope, for example where the ISO standards for air interface cover all the frequency range, EPC operates only within the UHF between 860-930MHz with one standard for 13.56MHz. The EPC standard activities, although taking advantage of the resources of the parent organisations in terms of expertise as well as potential users, is separate from the generic EAN UCC standardisation process. Such distinction is required due to the difference in the nature of standards and the need for a fast standard development process.

ISO vs. EPC

Whereas ISO can claim that it reflects the global requirements into a legitimate process (equal footing and consensus based), EPC focuses on speed and emphasises the broad support it receives from the industry community. The ISO and EPC processes can be seen as complementary, even more so when one consider that the only competing area is the standard for air interfaces frequencies. However, for both EPC supporters and for ISO the need for a single, global standard is impetuous. The benefits coming from standardization would be lost if in different parts of the globe, multinationals would have to invest in different technologies for RFID.

RFID STANDARDS USE

Technologically, RFID has the potential to simplify the process of tracking parts, without any line of sight and with multiple tags that can be detected simultaneously. Claimed benefits of RFID standardised technology supposed to improved supply chain efficiency, for example significantly lower transport and operating cost, reduced capital, or the stop of misplaced packaging during transport when moved between suppliers, or to cut off fraud.

The different actors in the RFID standards use are technology vendors, consultants and user companies. Whereas technology vendors proclaim RFID as a huge market opportunity to sell their technology and promise big benefits, the exact distribution of these benefits seems to turn out as big problem. For example, the standardised RFID technology seems not to be mature enough to satisfy the user requirements or integration with existing IT system. Additionally, in the case of IS implementation, the challenges associated with internal organizational change required by the change in the business processes due to RFID use create massive problems. Under these conditions, the users care less about standards and more about the practical cost-benefit analysis of the technology.

NEED OF RFID STANDARDS

The need for standards has become apparent to almost everyone. The existence of multiple variations of a symbology would make the job (manufacturing equipment) near impossible, without even thinking about the problems the end user would have. The explosive growth of barcode technology over the past ten years is due in part to the willingness of the various inventors of symbologies to put their inventions in the public domain and allow for open standards. However, all the RFID systems have one thing in common, they communicate over the airwaves. (4)

WHAT IS EPC AND WHY IS IT SO IMPORTANT?

EPC (electronic product code) is the RFID version of the UPC barcode standard. Like UPC, EPC is intended to be used for specific product identification as well as case and pallet identification. However, EPC goes beyond UPC by not only identifying the product as an SKU, but also providing access to additional data (via the EPC Network) about the origin and history of the specific units. The EPC tag itself identifies the manufacturer, product, version, and serial number. It's the serial number that takes EPC to the next level by providing the key to data related to specific lots/batches/units. This unit-level data is stored somewhere else (the internet or other network) but a standardized architecture allows to access the data much like you would access a web page (though this would be happening automatically behind the scenes). This architecture is known as the EPC Network.

EPC MYTHS AND MISCONCEPTIONS

EPC is strictly an RFID standard: Granted, RFID is part of EPC, but there is a lot more to EPC. Most notably is the EPC network, which is where all the data related to EPC will exist. This is a significant change in item-based data management and should not be taken lightly.

The use of RFID for EPC tags will allow them to hold more data than bar codes : This is simply not true. RFID tags could hold more data than a bar code, but under the EPC standard, they will not. So what kind of data will exist on the EPC network? Just about anything related to the item or container. The data flexibility is accomplished through the use of a new computer language called Physical Markup Language (PML) which is essentially a variation of the more commonly known Extensible Markup Language (XML).

RFID technology is capable of supply-chain management functionality, but the EPC standard is not utilizing it. Data will only be written to the tag once under the EPC standard. Any changes in status or other update information will be written to the EPC network, not the tag.

The use of RFID for EPC tags will allow them be more durable than bar codes : Probably not. While more expensive RFID tags encased in a plastic shell are more durable than bar codes in harsh environments, the lower cost unprotected RFID circuits glued to a paper label that are more likely to be used do not share these durability characteristics. These tags will prove be less durable than bar codes in many applications.

The current thrust of EPCglobal is known as UHF Generation 2 (UHF Gen 2), a Write Once Read Many tag with more memory (96 bits vs. 64 bits) than preceding Class 0 and Class 1 tags. UHF Gen 2 will also provide a bridge to the eventual Class 2 High Memory full Read Write tag. Prior to UHF Gen 2, Class 0 and Class 1 were being utilized for EPC, but they were not interoperable. Consequently, a retailer utilizing an EPC solution - such as Wal-Mart - would need different RFID readers to read different tags or force all of their suppliers into one technology. UHF Gen 2 will merge the Class 0 and Class 1 standards for a global, interoperable EPC standard.

EPC Tag Classes

Currently, several classes of tags fall under the EPCglobal umbrella. The difference between Class 0 and Class 1 is in the data structure and operation. Class 0 tags are read only. Class 1 tags are one-time writeable. The EPC standards call for 5 classes of tags over time. The following table outlines the roadmap for the EPC tag class type:

EPC Class Type	Features	Tag Type
Class 0	Read Only	Passive (64 bit only)
Class 1	Write Once, Read Many (WORM)	Passive (96 bit min.)
Class 2	(Gen2) Read/ Write	Passive (96 bit min.)
Class 3	Read/Write with battery power to enhance range	Semi-Active
Class 4	Read/Write active transmitter	Active

The chip manufacturer can only program the Class 0 tag; the Class 1 Version 1 tag can be programmed on the factory floor. While functionally equivalent under the EPCglobal classification system, Class 0 and Class 1 use different hardware technologies to implement the Identity tag functionality. Class 0 tags are programmed when they are manufactured (referred to as "Read-Only" or "R/O"), assuring uniqueness of the tag ID. Class 1 tags can be programmed once, referred to as "Write Once, Read Many" or "WORM", by the user, providing operational flexibility.

Class 0 and Class 1 tags also use different protocols, or "air interfaces" to communicate. So, while both Identity tag implementations perform the required functions, they cannot communicate with each other. Tags of both classes can coexist in an environment, but require readers that "speak their language" to be identified.

EPC code structure

The EPC is a number made up of a header and three sets of data. The header identifies the EPC's version number, allowing for different lengths or types of EPC later on.

* The second part of the number identifies the EPC Manager, most likely the manufacturer of the product.

* The third, called object class, refers to the exact type of product, most often the Stock Keeping Unit (SKU).
* The fourth is the serial number unique to the item. This makes it possible to quickly find products that might be nearing their expiration date.

FREQUENCIES IN THE PERSPECTIVE OF STANDARDS

In the UHF (Ultra High Frequency) band, where RFID tags work according to the principles of the electromagnetic coupling, the most popular technology at the moment is the one based on the ISO 18000-6C protocol, best known as EPC Class 1 Gen 2 or for short Gen 2. The EPC Class 1 Gen 2 standard was proposed by the private organization EPCGlobal and then adopted in 2006 as the ISO 18000-6C standard by the International Standards Organization (ISO).

The EPC Class 1 Gen 2 standard was created to address some issues of previous UHF RFID standards conceived for logistics applications (such as the ISO 18000-6a and the ISO 18000-6b). The new standard was developed specifically to track fluxes of goods between different companies and across all world regions with good read performance in environments with a high density of tags.

According to the standard specifications, EPC Class 1 Gen 2 tags have four memory banks: reserved, EPC, TID and user memory. The EPC bank, typically 96 bit in size, is the one that mainly characterizes EPC Gen 2 tags. It allows to univocally identify an enormous number of objects and controls anti-collision and wake-up functions. Since the EPC number is programmed by the user, more and more Gen 2 RFID tags in the market, as well as tags of other technologies, have a unique serial number that is set at the factory by the IC manufacturer and is inalterable in order to make the tag really unique. This feature is particularly important in applications where counterfeiting is an issue.

EPC Gen 2 RFID tags work in the frequency band that goes from the 860 MHz to the 950 MHz, but there are three main frequency sub-bands used in different geographical regions:

* Europe, India, Middle East, Africa: 865-868 MHz (ETSI)
* US (plus South America and some regions of Asia): 902-928 MHz (FCC)
* Japan: 950-956 MHz (JPN)

There are EPC Gen 2 tags that are designed to work well across the entire 860-950 band and others that are optimized to provide the best performance in a specific sub-band. The majority of the EPC Class 1 Gen 2 tags on the market are passive tags, but **Gen 2 BAP tags** and **semi-passive Gen 2 RFID tags** are also available.

Passive UHF Gen 2 RFID tags - Common advantages:

* Global functioning.
* Read range from a few centimeters (UHF near field) to more than 10 meters.
* Lower costs of labels in comparison to HF labels in good volumes.
* Good performance in data transmission and in environments with high tag density.

Passive UHF Gen 2 RFID tags - Common disadvantages:

* Difficulties with liquids (absorbing materials) and as well as in proximity to animals and the human body.
* Problems in the presence of metals (reflecting/ detuning materials) and when attached to metal surfaces if not specifically designed for this kind of usage.
* Smaller memory sizes in comparison to passive HF RFID tags.

BAP (Battery Assisted Passive) and semi-passive UHF EPC Gen 2 RFID tags - Common advantages in comparison with passive UHF Gen 2 RFID tags

* They offer longer read ranges (BAP tags, few dozens of meters).
* The battery power helps with liquids.
* Extra sensors (semi-passive RFID tags).

BAP and semi-passive UHF EPC Gen 2 RFID tags - Common disadvantages in comparison with Passive UHF Gen 2 RFID tags

* Higher costs than "comparable" passive technologies.
* The battery can limit their usage in some extreme environmental conditions.

Another RFID standard in the UHF frequency band is the **ISO-18000-7**, an active RFID protocol that works at 433 MHz that is also promoted by the DASH7 Alliance, a consortium that is working to create a new wireless sensor networking technology that has evolved from a combination of existing radio-frequency identification and sensing technologies.

ISO-18000-7 active RFID tags - Common advantages:

* Global functioning.
* Thanks to their working frequency, they have less problems with metals and liquids in comparison to UHF passive RFID tags.
* Long read ranges.
* Low battery consumption for longer durability.

ISO-18000-7 active RFID tags - Common disadvantages:

* As all tags with a battery, ISO-18000-7 RFID tags have higher maintenance costs than "comparable" passive technologies.
* The battery can limit their usage in some extreme

environmental conditions.

Like other active RFID tags that work at higher frequencies (usually at 2.45 GHz and 5.8 GHz) and are based on proprietary protocols, ISO 18000-7 RFID tags are used in indoor and outdoor Real Time Location Systems, in applications where long read ranges are required and where measurements of environmental conditions have to be made.

ISO 18000 series RFID standards

The ISO 18000 series standards is a series of standards that define the air interface for the different RFID frequencies in use around the globe. There is a total of seven standards withn the ISO 18000 series as outlined in the table below:

ISO 18000 Standard	Details of the particular ISO 18000 series standard
ISO 18000-V1	Generic parameters for air interfaces for globally accepted frequencies
ISO 18000-V2	Air interface for 135 KHz
ISO 18000-V3	Air interface for 13.56 MHz
ISO 18000-V4	Air interface for 2.45 GHz
ISO 18000-V5	Air interface for 5.8 GHz
ISO 18000-V6	Air interface for 860 MHz to 930 MHz
ISO 18000-V7	Air interface at 433.92 MHz

Why are retailers and consumer goods suppliers so interested in RFID?

The incremental gains by RFID are simply not that significant. The larger benefits to RFID are associated with a couple of issues that have been plaguing the retail and consumer goods industries for decades. That is, theft and product counterfeiting. Theft, in the form of shoplifting as well as larger scale theft occurring during transport and storage, is costing retailers and suppliers billions of dollars each year. On top of that, product counterfeiting is costing suppliers billions more. The obvious use of RFID would be to have readers at the exits (as is currently used for higher priced products) to catch shoplifters. The ability to quickly perform detailed reads of cargo as it moves through the supply chain is another way RFID could help to prevent theft. This would allow to more easily narrow down the source of the theft (to a physical location). In addition, it would make it much more difficult to sell the stolen product since it could be more easily identified (since each product would have a unique ID). Similarly, counterfeit products would be easier to identify. All would be needed is a portable computer and an RFID reader, and one could walk down the aisles of a suspect retailer and ID counterfeit or stolen merchandise.

ISSUES IN THE RFID STANDARDS

There are some issues in the RFID standards world. There are two real problems at present. Firstly there is a lack of participation by a number of vendor and user organisations in the ISO standards process. The second is that there is a lack of understanding about the procedures of the two main bodies involved in the specification of standards for RFID, namely ISO and EPCglobal. ISO was trying to move forward to support sensors with 18000-6C. Nothing else is happening on the sensor standardisation front with HF—where it should—because none of the technology vendors are participating! As a result, major applications for IATA and the automotive industry are at stalemate because they cannot get the chip that they need. In EPCglobal, vendors make IP aware to others taking part in the formation of a standard as part of their membership. The ownership of the IP is not transferred to EPC until the standard has been accepted by the membership. This means that EPC can't disclose anything to ISO until it has been accepted by its membership. ISO has some clearly stated and well-known rules around vendor IP that is put up for standards consideration, but EPCglobal developed their own rules.

The time has come for a better understanding of the roadmaps of each side. For instance, there is a working relationship between ISO and CEN[1]. For RFID, CEN/TC225 AIDC Technology has a responsibility to review various ISO standards and undertake its own developments. It needs to attract more users, vendors and consultancies to get involved. The problem is ISO is international; therefore the participants tend to belong to large companies, who are able to fund their employee's involvement. There is hope in Europe, as CEN committee is starting to look at European requirements and we could look to the EU to help funding for smaller organisations to be involved.

RFID: Update on Standards and Regulatory Initiatives

After extensive research, SC 31 (sub-committee) requires that all RFID standards include two clauses regarding RFID and health care. The first relates to RF emissions and the general population, specifically device manufacturers claiming conformance to this standard shall self-certify that RF emissions do not exceed the maximum permitted exposure limits recommended by either IEEE C95.1:2005 or ICNIRP according to IEC 62369-1. If a device manufacturer is unsure as to which recommendation to be cited for compliance the manufacturer shall self-certify to ICNIRP limits. The second addresses RF emissions and susceptibility in a health care setting, namely *Device manufacturers claiming*

---

1 CEN is the super-national standards body that sits above BSI, DIN (Germany) and AFNOR (France) and all the other European countries to provide harmonised standards right across all technologies. In French, CEN stands for Comité Européen de Normalisation (hence the letters CEN). In English the organisation is called European Committee for Standardization.

conformance to this standard shall self-certify that RF emissions and susceptibility comply with IEC 60601-1-2. Additionally, the RFID Experts Group, the Food and Drug Administration (FDA), and two universities (Georgia Tech and the University of Pittsburgh) are working together to address the three issues confronting RF emissions and human health:

* RF interference and the susceptibility of medical devices (implantables and wearables). This issue is addressed in the American Association of Medical Instrumentation's ANSI standard ANSI/AAMI PC69.
* RF interference and clinical instrument susceptibility. This issue is addressed in the International Electrotechnical Commissions (IEC) standard IEC 60601-1-2 and within a series of medical instrument standards covered in IEC 60601-2, and . . .
* The effects of RF on pharmaceuticals, biologics, blood products, human physiology, etc., addressed in ANSI/IEEE C95.1 and International Commission on Non-Ionizing Radiation Protection (ICNIRP) guideline, *Guidelines for Limiting Exposure to Time-Varying Electric, Magnetic, and Electromagnetic Fields (Up to 300 GHz).*

STANDARDS AND REGULATIONS

Moving into the area of standards and regulations there are several issues of which the audience needs to be aware. RFID should be operating in the Industrial, Scientific, and Medical (ISM) bands. The good news regarding ISM bands is that anyone can use them without a license; the bad news is that anyone can use them without a license. The ISM bands are the dominant frequencies for RFID, namely:

* 13.553–13.567 MHz (center frequency 13.560 MHz)
* 433.05–434.79 MHz (center frequency 433.92 MHz)

* 902–928 MHz (center frequency 915 MHz)
* 2.400–2.500 GHz (center frequency 2.450 GHz)
* 5.725–5.875 GHz (center frequency 5.800 GHz

There should be some amendments to provisions of the Radio Regulations necessary to provide for the operation of ship and port security and maritime safety systems.

In 2006 ISO TC 104/SC 4/WG 2 published ISO 18185-5, *Freight containers — Electronic seals — Part 5: Physical layer*, which required that all e-Seals support two frequencies (2450 MHz [ISO/IEC 24730-2] and 433.92 MHz [ISO/IEC 18000-7]). With this dual frequency requirement being placed ON THE TAG as opposed to the reader, plus the fact that these e-Seals are single use devices, the incremental cost of dual frequency e-Seals will be approximately 1 billion dollars per year. After the publication of 18185-5, there exists no e-Seal on the market or in R&D labs that supports this dual frequency concept. The maritime community is of the opinion that the solution for this issue is the allocation by the ITU-R for a maritime safety and security band that will support container security.

* ANSI MH 10/SC 8 is also working on a unique item identifier that will be compatible with the 96-bit EPC. Recommending ISO control of the remaining 240 headers of which EPC is a part, SC 8 will establish a Unique Digital Identification (UDI) also comprised of 96 bits. This 96-bit identification will be able to coexist with EPC and have the same performance benefits. Once completed within SC 8 the standard will be elevated to ISO.
* The IEEE standard 1451.7 (Sensors for RFID) is expected to be balloted within IEEE by the end of the year.

A SURVEY OF RFID STANDARDS

* AIAG ARF-1 „Application Standard for RFID Devices in the Automotive Industry".
* AIAG B-11 „Tire and Wheel Identification Label Standard"
* ANSI/INCITS 256 „Radio Frequency Identification (RFID)", NCITS 256 defines a standard for Radio Frequency Identification (RFID) for use in item management. This standard is intended to allow for compatibility and to encourage interoperability of products for the growing RFID market in the United States.
* ANS/INCITS 371: „Information Technology - Real Time Locating Systems (RTLS)."
    — Part-1: 2.4 GHz Air Interface Protocol
    — Part-2: 433 MHz Air Interface Protocol
    — Part-3: Application Programming Interface
* ANSI/MH 10.8.4 „RFID for Returnable Containers."
* AWWA IMT61457 „The Use of Mobile and RFID Data and Field Force Integration in a Major Water Utility"
* CEPT T/R 60-01: „Low-power radiolocation equipment for detecting movement and for alert" (*EAS*). Technical Recommendation.

    http://www.ero.dk
* CEPT T/R 22-04: „Harmonisation of frequency bands for Road Transport Information Systems (*RTI*)" (Mautsysteme, Frachtidentifikation). Technical Recommendation. http://www.ero.dk
* ECMA-340: siehe ISO/IEC 18092 (NFCIP-1)
* ECMA-352: siehe ISO/IEC 21481 (NFCIP-2)
* ECMA-356: siehe ISO/IEC 22536 (NFCIP-1; RF Interface Test Methods)
* ECMA-362 siehe ISO/IEC 23917 (NFCIP-2; Protocol Test Methods for NFC)

* EN 50061: „Sicherheit implantierbarer *Herzschrittmacher*". Vorschriften zum Schutz vor Fehlfunktion durch elektromagnetische Beeinflussung (entspricht VDE 0750). http://www.etsi.org
* EN 300 220: „Electromagnetic compatibility and Radio spectrum Matters (ERM); Short Range Devices (*SRD*); Radio equipment to be used in the 25 MHz to 1 000 MHz frequency range with power levels ranging up to 500 mW;" http://www.etsi.org
* Part-1: Technical characteristics and test methods
  — Part-2: Supplementary parameters not intended for conformity purposes
  — Part-3: Harmonized EN covering essential requirements under article 3.2 of the R&TTE Directive
* EN 300 330: „Electromagnetic compatibility and Radio spectrum Matters (ERM); Short Range Devices (*SRD*); Radio equipment in the frequency range 9 kHz to 25 MHz and inductive loop systems in the frequency range 9 kHz to 30 MHz"; http://www.etsi.org
  — Part-1: Technical characteristics and test methods
  — Part-2: Harmonized EN under article 3.2 of the R&TTE Directive
* EN 300 440: „Radio Equipment and Systems (RES); Short Range Devices, Technical characteristics and test methods for radio equipment to be used in the 1 GHz to 25 GHz frequency range with power levels ranging up to 500 mW"; http://www.etsi.org
* ETS 300 683: „Radio Equipment and Systems (RES); ElectroMagnetic Compatibility (EMC) standard for Short Range Devices (*SRD*) operating on frequencies between 9 kHz and 25 GHz"; http://www.etsi.org

* EN 300 761: „Electromagnetic compatibility and Radio spectrum Matters (ERM); Short Range Devices (SRD); *Automatic Vehicle Identification* (AVI) for *railways* operating in the 2,45 GHz frequency range"; http://www.etsi.org
    — Part-1: Technical characteristics and methods of measurement
    — Part-2: Harmonized standard covering essential requirements under article 3.2 of the R&TTE Directive
* EN 300 674: „Electromagnetic compatibility and radio spectrum matters (ERM); Road Transport and Traffic Telematics (RTTT); Technical characteristics and test methods for Dedicated Short Range Communications (DSRC) transmission equipment (500 kbit/s / 250 kbit/s) operating in the 5,8 GHz Industrial, Scientific and Medical (ISM) band"; http://www.etsi.org
* EN 301 489: „Electromagnetic compatibility and radio spectrum matters (ERM); ElectroMagnetic Compatibility (EMC) standard for radio equipment and services"; http://www.etsi.org
    — Part-1: Common technical requirements
    — Part-2: Specific requirements for radio paging equipement
    — Part-3: Specific requirements for Short Range Devices (SRD) operating on frequencies between 9 kHz and 25 GHz
    — Part-4: Specific requirements for fixed radio links and ancillary equipment and services
    — Part-5: Specific requirements for Private and Mobile Radio (PMR) and ancillary equipment (speech and non-speech)
    — Part-6: Specific conditions for Digital Enhanced Cordless Telecommunications (DECT)

equipment

— Part-7: Specific conditions for mobile and portable radio and ancillary equipment of digital cellular radio telecommunications systems (GSM and DCS)
— Part-8: Specific requirements for GSM base stations
— Part-9: Specific conditions for wireless microphones and similar Radio Frequency (RF) audio link equipment
— Part-10: Specific conditions for First (CT1 and CT1+) and Second Generation Cordless Telephone (CT2) equipment
— Part-11: Specific conditions for FM broadcasting transmitters
— Part-12: Specific conditions for Earth Stations operated in the frequency ranges between 4 GHz and 30 GHz in the Fixed Satellite Service (FSS)
— Part-13: Specific conditions for Citizens' Band (CB) radio and ancillary equipment (speech and non-speech)
— Part-15: Specific conditions for commercially available amateur radio equipment
— Part-16: Specific conditions for analogue cellular radio communications equipment, mobile and portable
— Part-17: Specific requirements for Wideband data and HIPERLAN
— Part-18: Specific requirements for Terrestrial Trunked Radio (TETRA)
— Part-19: Specific conditions for Receive Only Mobile Earth Stations (ROMES) operating in the 1,5 GHz band providing data communications

- — Part-20: Specific conditions for Mobile Earth Stations (MES) used in the Mobile Satellite Services (MSS)
- — Part-22: Specific requirements for VHF aeronautical mobile and fixed radios

* ERC/DEC 92-02: „CEPT/ERC Decision on the frequency bands to be designated for the coordinated introduction of Road Transport Telematic Systems"; http://www.ero.dk
* ERC/DEC 97-10: „CEPT/ERC Decision on the mutual recognition of conformity assessment procedures including marking of radio equipment and radio terminal equipment"; http://www.ero.dk
* ERC/DEC 01-01: „CEPT/ERC Decision: Non-specific short range devices in 6765 - 6795 kHz and 13.552 - 13.567 MHz"; http://www.ero.dk
* ERC/DEC 01-02: „CEPT/ERC Decision: Non-specific short range devices in 26.957 - 27.283 MHz"; http://www.ero.dk
* ERC/DEC 01-03: „CEPT/ERC Decision: Non-specific short range devices in 40.660 - 40.700 MHz"; http://www.ero.dk
* ERC/DEC 01-04: „CEPT/ERC Decision: Non-specific short range devices in 868.0 - 868.6 MHz, 868.7 - 869.2 MHz, 869.4 - 869.65 MHz, 869.7 - 870.0 MHz"; http://www.ero.dk
* ERC/DEC 01-05: „CEPT/ERC Decision: Non-specific short range devices in 2400 - 2483.5 MHz"; http://www.ero.dk
* ERC/DEC 01-13: „CEPT/ERC Decision: Short range devices for inductive applications in 9 - 59,750 kHz, 59 .750 - 60.250 kHz, 60.250 - 70 kHz, 70 - 119 kHz and 119 - 135 kHz"; http://www.ero.dk
* ERC/DEC 01-14: „CEPT/ERC Decision: Short range devices for inductive applications in 6765 - 6795

kHz, 13,553 - 13.567 MHz"; http://www.ero.dk

* ERC/DEC 01-15: „CEPT/ERC Decision: Short range devices for inductive applications in 7400 - 8800 kHz"; http://www.ero.dk
* ERC/DEC 01-16: „CEPT/ERC Decision: Short range devices for inductive applications in 26.957 - 27.283 MHz"; http://www.ero.dk
* ERC/REC 01-06: „CEPT/ERC Recommendation: Procedure for mutual recognition of type testing and type-approval for radio equipment"; http://www.ero.dk
* ERC/REC 70-03: „CEPT/ERC Recommendation 70-03 relating to the use of Short Range Devices (*SRD*)"; http://www.ero.dk
* ETSI TS 102 190: siehe ISO/IEC 18092 (NFCIP-1)
* ETSI TS 102 312: siehe ISO/IEC 21481 (NFCIP-2)
* ETSI TS 102 345: siehe ISO/IEC 22536 (NFCIP-1; RF Interface Test Methods)
* ISO/IEC 6346 „Freight containers - Coding, identification and marking"
* ISO/IEC 7810: „Identification cards - Physical characteristics"
* ISO/IEC 7816: „Identification cards - Integrated circuit(s) cards with contacts"
  — Part-1: Physical characteristics
  — Part-2: Dimensions and location of the contacts
  — Part-3: Electronic signals and transmission protocols
  — Part-4: Interindustry commands for interchange
  — Part-5: Registration system for applications in IC Cards
  — Part-6: Interindustry Data Elements
  — Part-7: Interindustry commands for Structured

Card Query Language (SCQL)

— Part-8: Security architecture and related interindustry commands
— Part-9: Enhanced interindustry commands
— Part-10: Electronic signals and answer to reset for synchronous cards
— Part-11: Card structure and enhanced functions for multi-application use
— Part-12: Cryptographic information application

* ISO/IEC 8824-1: „Information technology - Abstract Syntax Notation One (ASN.1) - Specification of basic notation."

* ISO/IEC 8825-1: „Information technology - ASN.1 encoding rules - Specification of Basic Encoding Rules (BER), Canonical Encoding Rules (CER) and Distinguished Encoding Rules (DER).

* ISO/IEC 9798: „Information technology - Security techniques - Entity authentication".

— Part-1: General
— Part-2: Mechanisms using symmetric encipherment algorithms
— Part-3: Mechanisms using digital signature techniques
— Part-4: Mechanisms using a cryptographic check functions
— Part-5: Mechanisms using zero knowledge techniques

* ISO/IEC 9834-1: 1993/Amd.2: 1988 „Information technology - Open Systems Interconnection - Procedures for the operation of OSI Registration Authorities: General Procedures."

* ISO/IEC 10373: „Identification Cards - Test methods".

— Part-1: General

- Part-2: Magnetic strip technologies
- Part-3: Integrated circuit cards (kontaktbehaftete Chipkarten)
- Part-4: Contactless integrated circuit cards (close-coupling)
- Part-5: Optical memory cards
- Part-6: Proximity cards (kontaktlose Chipkarten nach ISO/IEC 14443)
- Part-7: Vicinity cards (kontaktlose Chipkarten nch ISO/IEC 15693)

* ISO/IEC 10374: „Container - Automatische Identifizierung" („*Freight containers - Automatic identification*").
* ISO/IEC 10536: „Identification cards - Contactless integrated circuit(s) cards".
  - Part-1: Physical characteristics
  - Part-2: Dimensions and location of coupling areas
  - Part-3: Electronic signals and reset procedures
  - Part-4: Answer to reset and transmission protocols
* ISO/IEC 11784: „Radio-frequency *identification of animals* - code structure";
* ISO/IEC 11785: „Radio-frequency *identification of animals* - technical concept";
* ISO/IEC 14223: „Radio-frequency *identification of animals* - Advanced Transponders":
  - Part-1: Air Interface
  - Part-2: Code and command structure
* ISO/IEC 14443: "Identification cards - Proximity integrated circuit(s) cards":
  - Part-1: Physical characteristics
  - Part-2: Radio frequency interface

- Part-3: Initialization and anticollision
- Part-4: Transmission protocols

* ISO/IEC 14816: „Road Traffic and Transport Telematics - Automatic Vehicle and Equipment Identification - Numbering and Data Structures"
* ISO/IEC 15459: "Information technology - Automatic identification and data capture techniques - Unique identifiers for item management"
  - Part-1: Unique identification of transport units
  - Part-2: Registration procedures
  - Part-3: Common rules for unique identification
  - Part-4: Unique item identification for supply chain management
  - Part-5: Uniqe identification of returnable transport items (RTIs)
  - Part-6: Unique identification for product groupings in material lifecyle management
* ISO/IEC 15693: "Identification cards - contactless integrated circuit(s) cards - Vicinity Cards"
  - Part-1: Physical characteristics
  - Part-2: Air interface and initialisation
  - Part-3: Protocols
  - Part-4: Registration of Applications/issuers
* ISO/IEC 15961: "Information technology - RFID for *Item Management* - Data protocol: application interface".
* ISO/IEC 15962: "Information technology - RFID for *Item Management* - Data protocol: data encoding rules and logical memory functions".
* ISO/IEC 15963: "Unique Identification of RF tag and Registration Authority to manage the uniqueness".
  - Part-1: Numbering System
  - Part-2: Procedural Standard

- Part-3: Use of the unique identification of RF tag in the integrated circuit

* ISO/IEC 17358: "Supply chain application for RFID - Application requirements".
* ISO/IEC 17363: "Supply chain application for RFID - Freight containers".
* ISO/IEC 17364: "Supply chain application for RFID - Transport units".
* ISO/IEC 17365: "Supply chain application for RFID - Returnable transport items",
* ISO/IEC 17366: "Supply chain application for RFID - Product packaging".
* ISO/IEC 17367: "Supply chain application for RFID - Product tagging".
* ISO/IEC 18000: "RFID for *Item Management*: Air Interface".
  - Part-1: Generic Parameter for Air Interface Communication for Globally Accepted Frequencies
  - Part-2: Parameters for Air Interface Communication below 135 kHz
  - Part-3: Parameters for Air Interface Communication at 13.56 MHz
  - Part-4: Parameters for Air Interface Communication at 2.45 GHz
  - Part-5: Parameters for Air Interface Communication at 5.8 GHz
  - Part-6: Parameters for Air Interface Communication - UHF Frequency Band (868 / 915 MHz)
* ISO/IEC 18001: "Information technology - Radio frequency identification for item management - Application requirements profiles".

* ISO/IEC 18046: "RFID Tag and Interrogator Performance Test Methods".
* ISO/IEC 18047: "Information technology - Radio frequency identification device conformance test methods". Testmethoden für ISO/IEC 18000
  - Part-3: Test methods for air interface communications at 13.56 MHz
  - Part-4: Test methods for air interface communications at 2.45 GHz
  - Part-7: Test methods for air interface communications at 433 MHz
* ISO/IEC 18092: "Near Field Communication (NFC) Interface and Protocol-1 (NFCIP-1)."
* ISO/IEC 18185: "Freight containers - Radio frequency communication protocol for electronic seal".
  - Part-1: Communication protocol
  - Part-2: Application requirements
  - Part-3: Environmental characteristics
  - Part-4: Data protection
  - Part-5: Sensor interface
  - Part-6: Message sets for transfer btw. seal reader and host computer
  - Part-7: Physical layer
* ISO/IEC 19762: "Information technology AIDC techniques - Harmonized vocabulary".
  - Part-1: General terms relating to Automatic Identification and Data Capture (AIDC).
  - Part-2: Optically readable media (ORM).
  - Part-3: Radio frequency identification (RFID).
* ISO/IEC 21007: "Gas Cylinders - Identification and Marking Using Radio Frequency Identification Technology"

  — Part-1: Reference Architecture and Terminology
  — Part-2: Numbering Schemes for Radio Frequency

* ISO/IEC 21481: "Near Field Communication (NFC) Interface and Protocol-2 (NFCIP-2)."
* ISO/IEC 22536: "Near Field Communication (NFC) Interface and Protocol-1 (NFCIP-1); RF Interface Test Methods."
* ISO/IEC 23389: "Freight containers - read write radio frequency identification (RFID)".
* ISO/IEC 23917: "Near Field Communication (NFC) Interface and Protocol-2 (NFCIP-2); Protocol Test Methods for NFC."
* ISO/IEC 24710: "Information technology AIDC techniques - RFID for Item Management - ISO/IEC 18000 Air Interface Communications - Elementary Tag license-plate functionality for ISO/IEC 18000 air interface definitions".
* ISO/IEC 24729: "Information technology - Radio frequency identification for item management - Implementation guidelines."
  — Part-1: RFID-enabled labels and packaging
  — Part-2: Recyclability of RF tags
  — Part-3: RFID interrogator/antenna installation

The use of RFID has experienced a major kick start in the last 10 years and it is estimated that in another 10 the market will reach USD 27.59 billion - a development that would not be possible without the use of harmonized standards. The diversity of ISO standards is for RFID, highlighting some of the many areas reaping the benefits of their use: from libraries and gas cylinders to animal identification devices and cargo shipment. ISO standards contribute to a harmonized framework, improved transparency, efficiency and safety in a complex, but growing market, while optimizing business processes and reducing operational costs for companies. ISO's RFID standards also provide significant benefits in product tracking as goods move through the supply chain, including loss prevention, inventory control, and in-transit visibility. In response to the latest market trends ISO is currently working on standards that will take RFID identification to next level. In a world increasingly driven by e-commerce and online communication, International Standards help postal services to stay competitive: Globalization and growing economic integration is possible only with standards. There are standards on the use of:

* ISO 9001, the quality management standard, in supply chains
* ISO/IEC 17025, the laboratory accreditation standard, in a milk quality programme, run by Cornell University, covering 8 000 farms in the USA.

ISO/IEC 18000 - RFID AIR INTERFACE STANDARDS

ISO/IEC 18000 is a series of standards being created by ISO/IEC/JTC 1/SC31/WG4/SG3. They are creating RFID air interface standards for the item identification world. The ISO/IEC 18000 series of standards are currently in revision to make them more up-to-date and bring in new information. Each of the parts are at different stages. The updates are in one main area but also include other things as detailed below:

* A revision to all the parts of 18000 will include fixes to the standards based on actual issues discovered during the use of the standards along with the addition of the capabilities to use batteries and sensors with the existing technologies.

NEW ISO STANDARD FOR RFID

A new RFID standard has been announced by the International Organization for Standardization (ISO). The new standard is ISO 17367:2009 defines the basic features of RFID use in the supply chain, when applied to product tagging. This will help manufacturers and distributors to better track products and to securely manage traceability with RFID tags. ISO 17367:2009 makes specific recommendations with regards to encoded identification of the product; additional information about the product on the RF tag; semantics and data syntax to be used; data protocol to be used to interface with business applications and the RFID system; air interface standards between the reader and RF tag. The ISO has a much broader focus, and has created standards that relate to frequency spectrum allocation, physical dimensions for 'proximity' - about 10cm range, and 'vicinity' - about 50 to 70cm range - identification cards, animal identification, and item management. EPC Global has a very specific focus of developing standards for a system that would ultimately allow unique identification of (just about) every piece of goods manufactured on the planet, along with an information system that could retrieve a life-time history for such goods.

ISO 18186: RFID standard proposed by China's container

RFID standard has always been a sensitive topic, one on the standard on people do not consciously think of the "bottleneck" the word, RFID industry may be bigger, more output, the standard is not his own, always have to pay a heavy price, not to mention that RFID standard has been upgraded a few years for the high degree of competition for world leadership, should not be overlooked. At present, China's RFID standard from integration in the world is also a no small distance, which for the development of RFID industry with Chinese characteristics significantly negative. China's standardization work has been effective in promoting sustained, both at the national level or local level, standards have gradually increased awareness. It is reported that China has to bear 45 ISO / IEC technical body work of the Secretariat as of 24 ISO /

IEC technical body President and Vice-Chairman. Proposed or promoted by the development of China's international draft standard 210, which has been approved as international standards, up to 96. "Short step, no a thousand miles, no small streams converge into a Jianghai" standard-setting work in process, even a little bit of progress are to be encouraged by the RFID industry. Not long ago, China's RFID standards work has made a major breakthrough: a new container standard - "ISO / PAS 18186: container-RFID shipping label system" become ISO recognized norms of international public, This specification is networking in the field of logistics and material first proposed and actively promoted by the development of country, an official release from the ISO publicly available specification, using active RFID technology to achieve the container-related supply chain data Classification.

China is a container-producing countries, the World on 95% of the container produced in China, China Yong You container throughput, shipping volume, manufacture and handling equipment, the four "world first" fully capable of supporting the conditions in the industry to the standards set who made a bigger step. "ISO / PAS 18186: container-RFID shipping label system" comes from the ISO 17363 standard from the amendment, to the formation of the new ISO / NP18186, and then to ISO / PAS 18186, experienced experts have been revised to enhance the dust settles after, then, China's ability to develop international standards has increased by one level.

China GB standard container made: Developments

* July 1, 2007, ISO 17363 standard published in China's technical participation in the main unit of China International Marine Containers (Group) Co., Ltd., CIMC full participation in the drafting of this standard and technical discussions. ISO 17363 standard is the predecessor of ISO 18186 standards.

* March 10, 2008, from the Port Group Bao Qifan led the research team drafted a "container for supply chain monitoring application of electronic seals technical specifications" for the draft.
* Early April 2008 in Hamburg, Germany organized by the International Organization for Standardization (ISO) standards for meeting the container, wrap the sail to the General Assembly on behalf of China for the establishment of new international standards. At this meeting, the Port Group introduced the "E-box reusable container seal / reusable container eSeal" research and application of the relevant circumstances. A one-time sealed containers of electronic boxes and reusable, is reusable sealing industry to use electronic case there are too many problems, and finally the formation of ISO 18185 standard is a one-time e-box closure, no necessary to reconsider the case of reusable electronic seal.
* In April 2008 in Hamburg, Germany Conference "e-box reusable container closure" of the proposal was revised to "monitor the supply chain of container tag" proposal. The container supply chain management with RFID tags already have ISO 17363 standard. However, the standard is a joint project of TC104 and TC122, so TC 104 TC 122 also seek the views of the need to be each other's consent. Later, TC 104 and TC 122 wished to communicate, when amendments to this standard by the TC 104, and amendments to this standard solely to the TC 104, TC 122 for this request was refused. Finally TC 104 decided to revise this standard to apply the new standard number, that is ISO 18186 "logistics cargo shipping container-RFID tags Freight containers - RFID cargo shipment tag system", the standard amendments mainly by the Shanghai Port Group to take charge, China main aspects of the Beijing Institute of Shipping and the set of groups were involved.

* February 2009, as supplemented and amended, China submitted a new proposal. International Organization for Standardization (ISO) container Committee opened a "voting window", called on all members of the proposed development of China's new international standards for container tag proposal for a period of three months to vote.
* May 10, 2009, voting results, sponsored by the Shanghai representative of China to develop international standards for RFID container proposal was adopted in the German vote. According to the National Standardization Management Committee forwarded the ISO (the International Organization for Standardization) is the voting results show that, 14 countries agreed to the proposal, including Denmark, France, Germany, the United States, including the eight, together with China also expressed its willingness to participate Biaozhun making.
* July 2010, the International Organization for Standardization (ISO) officially launched the "ISO / PAS 18186: container-RFID shipping label system."
* August 9, 2010, the State Committee for Standardization, the Ministry of Transport jointly held a press conference in Beijing, announced that the Hong Kong group vice president Bao Qifan lead the development of the "container-RFID shipping label system" become International Organization for Standardization (ISO) recognized norms of international public.

The significance of RFID labeling containers

Flow of container, an important part of the process, has always been highly concerned about the port logistics industry. RFID as an automatic identification and short-range communication technology applications in containers mainly to solve several problems: First,

Container number of automatic identification of problems; second is the state of the container doors Automatic Identification and regulatory issues; Third container logistics and transport information in the supply chain automatic identification and management; also is intelligent and containers throughout the supply chain under the full visual problems. By addressing these issues, RFID technology (RFID) can support container transportation and logistics integration of information flow, information-tracking containers during transportation to eliminate the wrong boxes, drain boxes, container gate holes to improve the speed, real-time record container switching frequency and time, improve the efficiency of container transportation and security. RFID standard-setting on the container accelerated the development of standards for the right to compete for a major mission of the national discourse.

Conversely, the development of standards will also promote the development of the other three areas. In the field of container transport, standardization achieved such progress in the container on the result of China's power demand on RFID applications, RFID technology with independent intellectual property rights and the RFID industry chain matures. The criterion for the electronic tagging of the forward container will bring about effective long-term stability. The future battlefield of global RFID standards will usher in a more intense impact, many countries have secretly set out standards in the planning and schedule.

RFID STANDARDS TABULATION

The relevant ISO standards are:

The ISO/IEC 15459 series of standards for unique identifiers : Object identifier structures are defined within these standards for the ISO/IEC 15459 identifiers that can be unique to the individual data carrier. Other codes, for example batch numbers, can obviously be defined as unique data, but have to be treated as attribute codes within an RFID tag to avoid the risk of two tags having such a code as a unique identifier.

ISO/IEC 15418 which, in turn, points to the two major data dictionaries used in bar code applications. These are the GS1 Application Identifiers and the ANSI MH10 Data Identifiers. Each of these schemes already has a defined method for converting into an object identifier structure for encoding in an RFID tag and for the RFID communication chain.

ISO/IEC 15434, which defines a message structure which is highly suitable for a "read all" type of data capture environment. Some of the types of messages are themselves based on data dictionaries, which are potential candidates for being registered with the Registration Authority for RFID data constructs. The syntax of ISO/IEC 15434 does present challenges for direct encoding in an RFID tag. Some of the main RFID standards are summarised in the table below:

RFID Standard	Details
ISO 10536	ISO RFID standard for close coupled cards
ISO 11784	ISO RFID standard that defines the way in which data is structured on an RFID tag.
ISO 11785	ISO RFID standard that defines the air interface protocol.
ISO 14443	ISO RFID standard that provides the definitions for air interface protocol for RFID tags used in proximity systems - aimed for use with payment systems
ISO 15459	Unique identifiers for transport units (used in supply chain management)
ISO 15693	ISO RFID standard for use with what are termed vicinity cards
ISO 15961	ISO RFID standard for Item Management (includes application interface (part 1), registration of RFID data constructs (part 2), and RFID data constructs (part 3).

ISO 15962	ISO RFID standard for item management - data encoding rules and logical memory functions.
ISO 16963	ISO RFID standard for item management - unique identifier of RF tag.
ISO 18000	ISO RFID standard for the air interface for RFID frequencies around the globe
ISO 18001	RFID for item management - application requirements profiles.
ISO 18046	RFID tag and interrogator performance test methods.
ISO 18047	The ISO RFID standard that defines the testing including conformance testing of RFID tags and readers. This is split into several parts that mirror the parts for ISO 18000.
ISO 24710	Information technology, automatic identification and data capture techniques - RFID for item management - Elementary tag license plate functionality for ISO 18000 air interface.
ISO 24729	RFID implementation guidelines - part : RFID enabled labels; part 2: recyclability of RF tags; part 3: RFID interrogator / antenna installation.
ISO 24730	RFID real time locating system: Part 1: Application Programming Interface (API); Part 2: 2.4 GHz; Part 3: 433 MHz; Part 4: Global Locating Systems
ISO 24752	System management protocol for automatic identification and data capture using RFID
ISO 24753	Air interface commands for battery assist and sensor functionality
ISO 24769	Real Time Locating System (RTLS) device conformance test methods
ISO 24770	Real Time Locating System (RTLS) device performance test methods

RFID standards are now widespread in their use.

Although EPCglobal and ISO are separate organisations, there are efforts to move towards a single RFID standards scenario, rather than having two sets of competing RFID standards.

Complexities of RFID Standards

Standards are critical in RFID, be it payment systems or tracking goods in open supply chains. A great deal of work has been going on to develop standards for different RFID frequencies and applications. RFID standards deal with the following:-

Air Interface Protocol: The way tags and readers communicate

Data Content: Organizing of data

Conformance: Tests that products meet the standard

Applications: How applications are used

The way the world has gone about developing the standards is a bit complex. There are two major and somewhat conflicting organizations into the business - ISO and Auto-ID Centre (now handled by EPC Global). Tags are required to be disposable (manufacturer may not get the tags back from the retailer to reuse it). Hence, the primary mission for any standard developer is to make the tags low cost. It should operate in UHF, as only UHF delivers read range needed for supply chain applications. And since the goods are needed to be tracked as they move across the globe, the standards must be open and globally accepted. There should also be an accompanying network architecture, which would enable anyone to look up information associated with a serial number stored on a tag. The network too needs to be based on open standards.

EPC standards for tags are the class 0 and class 1 tags:

* **Class 1**: a simple, passive, read-only backscatter tag with one-time, field-programmable non-volatile memory.
* **Class 0**: read-only tag that was programmed at the time the microchip was made

Class 1 and Class 0 have a couple of shortcomings, in addition to the fact that they are not interoperable. One issue is that they are incompatible with ISO standards. The new EPC standard ~V Gen2 is designed to be fast tracked with ISO standards but for some disagreements over the 8 bit Application Family Identifier (AFI). ISO has developed RFID standards for automatic identification and item management. This standard, known as the ISO 18000 series, covers the air interface protocol for systems likely to be used to track goods in the supply chain. They cover the major frequencies used in RFID systems around the world.

Key relationships with other components

In order to understand the relationship with other RFID components, consideration needs to be given to three classes of data standard:

* Unique Item Identifiers which, irrespective of the data carrier technologies, are able to identify uniquely an item within the domain and scope of an application. This qualification about domain (or namespace) and scope is made because there are many false assumptions that any one scheme is the one that provides uniqueness, and that uniqueness is assumed to be a timeless parameter.
* Data dictionary standards that define individual data elements, which have specific meaning within application domains.
* Message standards, particularly with respect to their role of identifying primary data dictionaries.

DATA STANDARDS: INTERRELATIONSHIPS

The Registration Authority is fairly key to this process. Although the flow appears to be through the Registration Authority, it is actually using code structure registered with the Registration Authority to achieve data presentable in an RFID format. To be compliant with ISO standards, the RFID format needs to be based on object identifiers, which ensures uniqueness from the business application through to the tag. The use of object identifiers in RFID applies to all the legacy types of data, including the unique item identifier. It even applies to new classes of unique item identifiers. There is one exception to this rule, which is the set of serialised EPCglobal codes. These codes can be encoded directly on the RFID tag, but this approach carries a penalty. By being the exception to the rule only tags that have specific architectural structures can support the EPCglobal code structure without it being linked to an object identifier structure. An interesting future challenge is how EPCglobal addresses the encoding of object-related data such as batch number and expiry date, which are strongly supported along with many other data elements by the well-established GS1 Application Identifier dictionary.

EPCGLOBAL STANDARDS

The relevant EPCglobal standard is the EPC Tag Data Standard (TDS). The current published version only specifies the serialised codes, i.e. those that are unique identifiers.

Significant development areas

One of the early challenges for JTC1 SC31 was how to identify data in a structured manner that would be invisible to the human eye. This was necessary to ensure that only tags that belonged to a specific domain were read. Other data would need to be ignored and excluded from the domain at the earliest possibility. None of this was possible with the closed systems applications for RFID, and data could be transferred from the interrogator to the application and only then identified as not being part of a domain. There was always the statistical risk that some false data would be ambiguous and be accepted.

By adopting an object identifier structure, every individual piece of data encoded on an RFID tag can be identified and arranged in whatever logical sequence made sense. The rules for object identifiers are specified in ISO/IEC 9834-1 and, by their very nature, object identifiers can be extremely long. The advantage they offer of being applicable to any type of data including the extensive legacy data already supported by bar code technology, had to be balanced by some efficient means of encoding on the RFID tag. Discussions with the relevant ISO committee resulted in a new structure for an object identifier, basically partitioning it into two parts:

* A Root-OID, that identifies all the common arcs of the object identifiers encoded on the RFID tag.
* A significantly shorter Relative-OID for the remaining part of the object identifier, with only this part encoded on the RFID tag.

The Root-OID can be implied, but to ensure interoperability of standards, part of the work of the Registration Authority is to assign a Data Format that is encoded on the tag and declares the Root-OID. This approach enables efficient encoding on the RFID tag, but all the communications from the device interface upwards can be based on the full object identifier to avoid any ambiguity within the systems. One of the challenges of encoding legacy data is that this is sometimes long and structured with punctuation and other characters for ease of reading by humans. There are two reasons why this presents a problem for RFID:

* The first, and most serious, one is that a number of tags have a limited encoding capacity (particularly

for the UII) and many of the legacy codes are too long to be encoded.

* Even if a tag has the capability of encoding a long legacy code, this still requires more air interface transmission time and if the requirement is to read multiple tags quickly, this can also present a problem.

There are two Registration Authorities that are key to extending the capabilities of RFID to support different applications:

* The Registration Authority for ISO/IEC 15961-2 is key to ensuring that object identifiers and other "data constructs" are registered. This allows RFID tags to be used in a reasonably generic manner, independently of the data content. In addition, the Registration Authority registers mapping tables that are used to achieve greater encoding efficiency on the RFID tag. These mapping tables have been designed to be machine readable, so that they can be embedded in interrogators to increase the level of automation and to rapidly support new applications as they are developed.
* The Registration Authority for ISO/IEC 15459 is also important, because one of the key features with many RFID tags is being able to - even requiring - the encoding of a unique item identifier. This particular register has enough capacity to be able to identify virtually any kind of object from any reasonable sized domain.

At present, even the existence of these Registration Authorities is known to very few organisations. Some responsibility rests with the vendor community to promulgate information about the benefits and issues. The registers2 can provide an important source of information for achieving higher levels of interoperability.

STANDARDS—WHAT GOOD ARE THEY?

Doing things according to standards—or accepted rules of conduct (behavior)—is critical for everyday life. Without adherence to these norms, automobile traffic flow through intersections could be deadly, and people wouldn't be able to communicate using telephones, faxes or the Internet. The accepted rules of conduct for RFID are embodied in such standards as ISO/IEC 18000-6 Type C, EPCglobal Gen 2 UHF, ISO/IEC 15961, ISO/IEC 15962, EPCglobal TDS 1.5, ISO 1736x and numerous others. This partial list of relevant standards illustrates what is perceived as the problem—there are many standards, and they can be very difficult to comprehend. It is especially difficult for those who are not RFID experts. Trying to sort through, decipher and understand the content of all of these standards—and then to implement them—can be very daunting. The purpose of B-11 Rev 8 is to enable those who don't want to become experts in a new technology to still be able to use it, and to understand it at a high level without having to master the nitty-gritty.

RFID Applications—Closed-loop / Limited Application

Most of these programs are closed-loop systems, operating either within a single organization or between two parties. Closed-loop in this context means that the information being carried by the RFID tag is retained, and has meaning only within the company or organization implementing the RFID system. Such systems can be implemented without complying with any standard regarding data format or the meaning of such information. Those who implement the system know how to encode and decode the data in a tag, and that data only has meaning to those specific entities.

Another observation is that most RFID implementations overwhelmingly place onto the tag and use only one piece of data. That single bit of information is the unique item identification (UII) number, or the "name" of the tagged item. These implementations do not make use of a powerful capability of RFID—the ability to store additional information on a tag beyond just the UII. While utilizing just the UII can be effective, there are numerous instances—and, in fact, entire industries—in which the use of additional data on a tag provides considerably more value. In some cases, this additional data provides the cost-benefit tipping point for using RFID.

RFID Application—Open-loop / Full-featured

The automotive original equipment manufacturing (OEM) industry has, for many decades, routinely exchanged parts and data through multiple parties and several levels within a supply chain. Today, identifying and tracking all of these parts—in real or near-real time—is a monumental task, and various means of automatic identification are currently employed. RFID is among them, and is being used precisely for two reasons:

* It can provide more than a simple name for an object.
* The additional data in memory can be changed during the part's lifespan.

One of the biggest stumbling blocks to widespread RFID implementation appears to be not only identifying which RFID standards are relevant, but also understanding those standards well enough to be able to implement them. The good news is that the International Organization for Standardization (ISO) and EPCglobal have effective RFID standards. The better news is that they work well together. In fact, in critical areas, relevant portions of these standards are identical! Most organizations that want to employ RFID do not have the requisite knowledge. Ideally, they would like to purchase what they need, and then be off and running. Unfortunately, the supplier base is not quite yet at that level, so those organizations that want or need to delve deeper into the technology—and then implement it—must struggle to understand the standards on their own.

The AIAG has successfully created effective, implementable standards for more than 25 years. The B-11 Revision 8 standard provides detailed, easy-to-understand explanations, with examples, of the relevant parts of the ISO and EPCglobal standards required to implement full-function RFID identification and tag-data storage systems. The overall goal of B-11 Rev 8 is to show that common data needs (unique item identity, plus additional data about that item) can be satisfied using globally applicable data syntax and implementation standards—regardless of what is carrying the information.

AIAG B-11 REV 8

Automatic identification and data collection (AIDC) technologies have been critical to the automotive industry's day-to-day processes for many years. These technologies include linear bar codes (Code 39, Code 128), two-dimensional symbologies (QR, Data Matrix), direct part marking (engraving, etching and peening) and RFID. One critical factor in successful RFID implementations is to make its use transparent with respect to the above-mentioned AIDC methods. The goal is to make the data the important part, and to make the data carrier device irrelevant, whether it's a bar code or RFID tag. To do this, a standard way of creating the information and putting it on the data carrier is needed—one song, with everybody knowing the notes, but playing them on different instruments.

The B-11 Item-Level RFID Standard was created in 2001 as a solution to one very specific problem—namely, how to track tires. Since 2001, with global input, B-11 has grown to cover not only tires, but any item or object. Revision 8 of the standard explains the global answer to a global question: How do we use RFID to carry data interchangeably and seamlessly with other AIDC methods?

Specifics

There are four data blocks on an ISO/IEC 18000-6 Type C / UHF EPC Gen 2 RFID tag: MB00, MB01, MB10 and MB11. (The numbers are actually binary numbers and are usually followed by a subscript 2; MB00 is the first memory block, and MB11 is the fourth memory block.) To maximize the benefits, there are two data blocks that need to be used *together* when storing various data elements: the Unique Item Identification Memory Block (MB01, the second memory block) and the User Memory Block (MB11, the fourth memory block). An item's unique item identification, or name, is placed into MB01. Additional data about that item is placed into MB11.

The ISO and EPCglobal RFID standards have very precise rules for creating the UII that would be placed into MB01. They are unambiguous, each method resulting in a unique name. They are also complex. In the case of EPCglobal-based structures, any of the identity types as described in the published TDS 1.4 or draft TDS 1.5 could be used as a name. For ISO-based structures, the use of the data identifier "25S" as the name serves the same purpose. The B-11 Rev 8 standard contains details on and examples of both schemes. There are also several methodologies of encoding the additional data into MB11, outlined in the ISO and EPCglobal standards. Again, B-11 Rev 8 provides details and examples.

The AIAG B-11 Rev 8 standard was created to provide easy-to-understand guidance on a difficult subject, through unambiguous explanations and examples. While there are many different RFID standards that need to be understood, the detailed examples contained within the B-11 provide guidance that can assist in enabling and easing future RFID implementations. There is now no reason why radio frequency identification cannot be rolled out to the global using-community in open-system, standards-based, multi-partner supply chains. Revision 8 of the AIAG B-11 Item-Level RFID standard, published on Nov. 16, can be obtained by contacting AIAG at (248) 358-3303, or by visiting the group's Web site, www.aiag.org.

Tag Memory Banks—MB01 and MB11

The RFID tag being discussed is one that complies with the ISO/IEC 18000-6C / EPCglobal UHF Gen 2 air interface protocol. Such tags have four memory banks, two of which store a user's data on the tag. These two locations are technically referred to as memory banks 01 and 11, often abbreviated as MB01 and MB11 (pronounced "em bee zero one" and "em bee one one"). The size of MB01 is usually very limited; its sole purpose is to contain the name by which the tag (as well as the item to which it is associated) is known. MB11's capacity is generally much larger, and is the place where all additional user-defined data will be stored.

Memory Bank 01—The "Name"

Memory Bank 01 contains the information referred to as the Unique Item Identification (UII) number. This piece of data is created at the time that the tag is placed onto an item, and can be thought of as that item's birth record, or name. This name must be unique in the world. This uniqueness is one of the desirable features of RFID technology, and is assured if one correctly follows the published standards. Once this UII has been created and written into MB01, there is great benefit in locking this data to prevent future changes. Locking the UII data in MB01 will not affect user data that may be written into MB11. The organization that places a tag onto an item has the responsibility of creating the UII and ensuring that it is a unique name. Various standards organizations have defined rules for not only creating the name, but making it unique in the world—if the rules are strictly followed.

Memory Bank 11—The "Data"

Memory Bank 11 is also called the user data memory

bank, and the information it contains will be referred to here simply as the data. As is possible for any -6C memory bank (that isn't locked), information placed into this memory location can be written, read and rewritten. There are a number of methods available for encoding and decoding the data bits comprising the information placed into MB11. As declared by the AIAG B-11 standard, all of the data stored within MB11 must include a data identifier for each data item. Internationally accepted data identifiers can be found on a list maintained by AutoID.org. Should the need arise for a new data identifier, the MH10.8.2 document found at the above link contains a request form and procedure for creating a new data identifier to meet the needs indicated by the requestor. Using these published data identifiers will enable reader and software vendors to create applications that can read tag data and decipher its meaning. This is absolutely necessary for true open-loop RFID applications.

10

Applications of RFID and Vertical Industries

ITS HERE! ITS THERE TOO!

Radio frequency identification technology (RFID) has begun to revolutionize the way business operations are done. Already, the U.S. Department of Defense and Wal-Mart have required their top suppliers to have RFID technology implemented in their supply chains. The technology, first invented by Harry Stockman with the publishing of his 1948 report "Communication by Means of Reflected Power," has only in the last few years become viable. As more manufacturers of RFID tags, readers and software solutions enter the market, the price of implementing the technology will continue to decline. The key feature of the technology is the ability for an RFID-tagged object to be tracked instantly from anywhere in the world, provided that a reader is in range. The main application for this today is tracking products along a supply chain. Everyone in the RFID-enabled supply chain has the ability to instantly call up the location, condition, and supply of a particular product. Industry professionals expect the RFID tag to eventually replace the barcode as the identification system of choice. And there are several reasons why this should eventually happen.

First, however, it is important to understand how the technology works. There are two types of RFID tags. An **active tag** uses its own battery power to contact the reader. It works over a greater distance than passive tags, but its larger size is its main drawback. On the other hand, a passive tag does not require a battery. A **passive tag** derives its power from the electromagnetic field created by the signal from the RFID reader. This generates enough power for the tag to respond to the reader with its information. While the range is smaller than active tags, it has no need for a battery. It makes the passive tag's life virtually unlimited and its size potentially minute.

In contrast to the barcode, RFID tags can be read from a larger distance, up to nine meters with some readers. Also, the RFID chip can be read from a reader not in its line of sight. For instance, cargo within a truck does not need to be unloaded to be scanned by an RFID reader, as the radio waves can penetrate the walls of the truck. An RFID reader can read hundreds of chips virtually instantaneously, speeding up the loading and delivery processes. Furthermore, some RFID tags contain other sensors that can transmit valuable data to managers.

THE VALUE PROPOSITION OF RFID

The basic but critical elements of an RFID system include tags, printer/encoders, reader/encoders (interrogators), sensors, middleware (for data-filtering and data-flow management). Some software adaptations may be required to enable legacy applications and systems to receive RFID-generated data. One really needs to look at this on a case-by-case basis to get at specific benefits. There is a need for committing to a pilot to create a robust business case for utilizing this technology. Due to automated reporting of real-time, accurate data can provide tremendous advantages in all kinds of industries. In addition, it's ability to support asset tracking and the creation of a pedigree is of great interest to a number of industries. Various features can be implemented into the RFID software to provide the flexibility needed for an individual requirement:

* Fully automated shipping and receiving
* Real time inventory
* Automatically order products that are low in inventory
* Real time security
* Real time to generate shipping and receiving documents
* Real time marketing information about different brand of products

APPLICATIONS, UTILITIES AND USAGES

RFID has many applications. It is often described as a transformational technology in terms of its potential impact on business processes and systems. It is a logical evolutionary step on from the barcode as a way of gaining increased labour productivity through automation. It can remotely sense objects to determine their identity track their position and detect properties such as pressure and temperature, when used in conjunction with allied technologies. With the ability to store several k bytes of data, it could be viewed as a form of 'mass distributed database' that has the potential to become ubiquitous - billions of tags in daily use throughout the world on all objects that are produced, stored, moved, sold and maintained.

There are numerous potential applications of RFID. Some are:

* Providing markings for electronic equipment, automobiles, and other large items with unique

serial numbers or codes. By scanning the RFID, the item can be located by its unique serial number.

* Marking shipping containers, pallets, or cartons. This application is being implemented in the retail industry by Wal-Mart, Target, and others.
* Attaching medical history to a person so that during an emergency an EMT or other medical personnel could quickly gain access to the patient's history.

For example, Wal-Mart was paving the way with a test program for RFID that the company hopes will eventually replace bar codes. In early 2004, they began testing with eight manufacturers and one distribution center. The manufacturers implemented case and pallet level RFID tags. In January 2005, Wal-Mart went live with RFID at distribution centers in Dallas. The initial 100 suppliers have since grown to more than three times that number to 500 stores and five distribution centers.

1. Common Uses of RFID

RFID systems can be used just about anywhere, where a unique identification system is needed. The tag can carry simple as well as complex information. Here are a few examples of how RFID technology is being used in everyday places:

1. RFID in retail stores offer real-time inventory tracking that allows companies to monitor and control inventory supply at all times.
2. RFID systems are being used in hospitals.
3. RFID chips for animals are extremely small devices injected via syringe under skin. Under a government initiative to control rabies, all Portuguese dogs must be RFID tagged by 2007. When scanned the tag can provide information relevant to the dog's history and its owner's information.

More practical uses of RFID tags have emerged as well:

* In the pharmaceutical industry, RFID tags on drug bottles are being used as anti-counterfeiting devices.
* Pet owners have begun implanting their cats and dogs with RFID chips to locate them should they become lost.
* In libraries, books are being tagged for self-automated checkout, freeing up librarians to perform other tasks. This also allows a librarian to easily locate a book misplaced on the wrong shelf.

2. Tracking and Locating Ability

* **Assets**
  - Find mobile equipment quickly
  - Improve utilization
  - Reduce loss and theft
  - Improve management visibility and control
* **Stock**
  - Perform real-time stock checks
  - Locate missing stock quickly
  - Comply with process and movement history rules
  - Increase business with improved customer response and availability
* **Staff**
  - Locate key workers quickly
  - Increase workforce productivity
  - Optimize mobile resources
  - Protect personnel and loan workers
* **Indoor RFID Tracking**

Passive RFID systems can be used over short-ranges where the objects to be tracked pass through a small number of known narrow points that can be automatically monitored - such as a pallet of goods passing through a door in a warehouse. Passive RFID systems are only capable of recording the movement of objects at those specific choke points and loose visibility outside of those points.

Active RFID provides superior performance where real-time tracking is required throughout a larger or more complex area, such as within a hospital, industrial or office building with many corridors, doors and rooms. Battery powered active tags have far greater signal range and the active readers are generally cheaper, thus making deployment more cost effective overall. An active RFID system is viable where the objects to be tracked are higher value, mission critical, have a consequential impact if not quickly located, or there is an additional security, regulatory or health and safety requirement for such information.

Tagging gives customers the ability to manage the following:

* Asset and Item Protection
* People and Asset Visibility
* Dwell Time or Time and Attendance
* Access Control
* Mobile Resource Optimization
* Evacuation Management
* Area Detection

Organizations will achieve the following benefits by using Tagging:

* Protect personnel
* Secure facilities
* Locate assets/equipment
* Optimize mobile resources
* Reduce asset loss
* Increase workforce productivity
* Perform real-time stock

* Improve estate utilization
* Compliance to process

3. RFID Solutions for ID Badges and Access Control

Most organizations today require a performing and efficient access control system. There are many reasons for having such a system and these include:

* Enhanced security to limit access to restricted areas,
* Tracking employee activity,
* Improve loss prevention and
* Compliance to higher internal or government regulated security measures

RFID technology is quickly becoming a popular choice since the advancement in technology makes a variety of solutions available to any organization. There are major deficiencies in the legacy solutions that use the traditional access control systems. Barcodes, magnetic stripes, and proximity readers all rely on the user to make contact or place the badge very close to the reader. Bar codes can only be read one at a time and the respective embedded information cannot be updated. Such limitations can be cumbersome and time consuming.

Benefit of using RFID : An RFID access control system can provide an easy and efficient solution. RFID badges can be read from much further distances than other traditional technologies and the embedded electronic information for each badge can be over-written repeatedly. Furthermore, multiple RFID badges can be read With conventional systems, user must handle the badge and place it close to (7-30 centimeters; 3-12 inches), or make physical contact with the reader. Few RFID access control technology allow the user to enjoy complete hands free access control. Few access badge usually eliminates the need to handle the badge or walk very close to the reader. To make installation simple, many models of RFID readers can connect to the access control computer through wireless communications. This eliminates the need for long wiring runs. It also allows a reader to be easily re-positioned or moved as needs change. The RFID access control system can report any unauthorized access and issue an alert to the host software.

RFID solve the access control concerns of an organization. Innovative RFID solutions helps a company improve security and make employee activity tracking an easy and efficient task. Few systems can be designed to ensure that an organization is compliant with the latest government security regulations. The RFID offers a variety of cards and readers that either fit perfectly in an organization's environment

4. Inventory control

The movement and tracking of goods through the manufacturing and supply chain process is a complex procedure which is difficult to manage. For example, sometime the goods being distributed to the retailer must go through one or more third party distribution processes, before they reach their final destination. Currently, most material tracking systems employ two-dimensional barcodes that must be close to and within the "line of sight" of the barcode reader. This requires manual scanning or a conveyor-like process to position the barcode and scanner. The scheduled scanning or manual methods cannot ensure the inventory remains up-to-date, due to oversights, errors, and internal shrinkage. With RFID solution, inventory can be updated in real time without product movement, scanning or human involvement. The system could also trigger automatic orders for products that are low in inventory.

* Benefits of using RFID Systems:
  - Provides total asset visibility
  - Reduces shelf space
  - Provides higher-level security

- Reduces errors
- Gives full inventory history
- Allows reduced inventory-stocking levels
- Facilitates "Just-in-Time" deliveries
- Provides full process control for products in the facility
- Reduces lead-time
- Shortens cross docking time
- Speeds up sort/pick rate
- Reduces overall cost of operations

5. Manufacturing Controls

Manufacturers can use RFID tags to keep track of Work-In-Process (WIP) inventory by affixing inexpensive passive RFID tags directly to the products being manufactured, or to bins that hold the products. Tags can contain only 'license plate' information used to uniquely identify each product, or they can contain a complete manufacturing history of the product which is updated at every step of manufacture - this would be beneficial if reliable access to the corporate network is not available at all stages of the production process. Specialized RFID tags exist that can withstand harsh manufacturing conditions, including extreme heat, corrosive liquids, and vibration. Some specialized Low Frequency (LF) tags can even be embedded *within* metal. The main obstacle to the spread of RFID in manufacturing is the success of barcodes, the previous generation of auto-ID technology. In most applications, reading barcodes requires human intervention, a clean, high-contrast environment, and often more than one attempt. In addition, not only is the amount of data that can be stored in a bar code is much smaller than in an RFID tag but it cannot be updated. In many applications the functional potential of RFID may not yet beat the low cost of barcodes, but it is only a matter of time before it does.

Manufacturers can track and record in-process assembly information into the RFID tag as an item progresses along the line.

Certain RFID solutions are ideal for manufacturers who build several products on a single production line, or manufacture complex or customized products. Assembly line personnel could use an RFID reader to verify which processes have been completed, to determine which inspections or tests are required, production planners and inventory control personnel could use the RFID tags to automatically update the customer database and finished goods inventory, using an RFID reader and PC. Benefits of using RFID Systems in manufacturing are as following:

* Maintains current item information on the tag - ideal for managing production of complex or customized products and assemblies, eliminates the need for separate paperwork on assembly status and content
* Can automatically notify the central product database when each process has been completed
* Real-time tracking of inventory, part kits and sub-assemblies

6. RFID Applications in Supply Chain Management

There are as many RFID applications as there are businesses. Some RFID applications have been around for decades, others are just starting to emerge. The roots of RFID technology can be traced back, (the 1970s) when the New York Port Authority introduced an RFID device used for toll collection. Today, the largest RFID application aids companies and governments in supply chain management. RFID is being used to manage products through production, distribution and retail. Experts say that early adaptors of RFID applications in supply chains have seen a significant increase in revenue.

7. RFID Applications in the Distribution Chain

RFID Applications can be used to monitor and

manage the movement of the finished products throughout a supply chain. RFID tags can be attached directly to the items and materials or they can be attached to the containers that carry them. Pallets, trailers, totes, carts, cargo containers, and reusable transport items can all be tagged. Readers placed throughout a facility can monitor movement and location of inventory, thus providing real time data. RFID applications in the supply chain enable more frequent and accurate inventory counts RFID applications in the supply chain can also decrease costs associated with inventory counting.

***RFID Applications in Retail and Product Marketing*:** RFID application in retail can greatly aid in reducing the cost of keeping accurate inventory data. With fewer people and less time, retailers can keep accurate inventories. RFID applications are, also, a significant aid in deterring theft in retail environments. RFID applications have been successfully deployed for anti-theft purposes for several decades.

***RFID Applications in Brand Protection*:** RFID applications can be used to protect brand identity by product marketing managers and to protect consumers from counterfeited products. RFID applications in anti-counterfeiting are, generally, aimed at high-value end consumer products. The unique identification number on standard RFID tags can be used to verify the authenticity of the products to which they are attached. These unique numbers combined with encryption algorithms can create authentication schemes that are extremely difficult for counterfeiters to circumvent.

RFID applications in anti-counterfeiting can be used in pharmaceuticals, electronics, entertainment, retail, IT and many more industries. RFID applications represent one of the most promising methods for countering this major problem. Manufacturers can trace products throughout the supply chain. The pharmaceutical industry is creating an RFID application that will document the authenticity of their products at retail.

By placing RFID tags on products at the point of manufacture, not only can products be traced throughout the supply chain, but it can also prevent counterfeit products from entering into the supply chain. RFID application in the supply chain offer solutions when it is impractical to use other technologies or manual labor to collect data. Data can provide suppliers and end users with the tools and information to make management decisions.

8. RFID In Hospitals

An RFID system can be used to track patients, doctors and expensive equipment in hospitals in real time. Few RFID technologies can also provide an electronic link for wirelessly communicating patient data. These applications can be combined with access control to allow only authorized personnel to access to critical areas of the hospital. Benefits of using RFID Systems in hospital environment are as following:

* Continuously track each patient's location
* Track the location of doctors and nurses in the hospital
* Track the location of expensive and critical instruments and equipment
* Restrict access to drugs, pediatrics, and other high-threat areas to authorized staff
* Facilitate triage processes by restricting access to authorized staff and "approved" patients during medical emergencies, epidemics, terrorist threats, and other times when demands could threaten the hospital's ability to effectively deliver services

9. RFID Enabled Wristbands

In simple terms, RFID wristband systems contain two parts: a tag and a reader. An RFID tag embedded in the wristband contains an antenna; the majority of these tags are passive. Passive RFID tags are small data transponders that are powered by the radiation received from a querying RFID reader – these tags "wake – up" when irradiated by the reader. Active RFID wristbands contain batteries and can send signals to readers. Readers can be mounted on walls, doorways, ceiling or be hand held.

High read reliability UHF Wristband: This unique design, which uses multiple inlays in a single band, is designed to ensure highly reliable reads in the chaos of disaster evacuation situations. The band is designed for long life in the difficult evacuation environment. In addition, few vendors provides a pre-perso service to the system integrator, linking the RFID data to the human and machine readable data to provide highly reliable correlation of the wristband wearer to their ID.

General Purpose UHF Wristband: RCD leveraged its extensive experience in high reliability wristband inlays to develop a highly cost effective UHF wristband that can be applied in a wide range of applications. With excellent read range, tolerance of a wide range of environmental conditions, and modest cost, this band is suitable for personal ID and people tracking applications in theme parks, cruise ships, conventions, hospitals and institutions.

RFID Wristbands in Healthcare: There are many applications for RFID technology in healthcare. These applications can provide patient tracking and identification services through the use of RFID wristbands. Additionally the use of RFID wristbands in healthcare can improve resource utilization and expedite patient admission process within all hospital departments. In addition, RFID wristbands can be used to record patient information to correctly identify patients. Incorrect identification of patients can lead to incorrect drug administration, execution of undesired procedures, misfiling of results and medical documentation and can cause general delays. With RFID enabled wristbands patient information, can be retrieved from the database via a wireless network by getting RFID enabled handheld devices in the vicinity of the wristbands.

RFID Enabled Wristbands in the Entertainment Industry: RFID wristbands can have various applications in the entertainment industry. Amusement parks, cruises, concerts and festivals could all benefit from RFID wristband deployment. RFID wristbands in amusement parks can provide customers with convenience and security. With the use of RFID wristbands parks can prevent children from being lost. RFID wristbands assigned to park goers at the point of ticket purchase, can store basic information such as name, telephone number and address of the person wearing the wristband. Parents who lose their children can go to any service kiosk, wave their own wristband, and bring up maps showing their kids' locations.

RFID Wristband Application in Defense : RFID enabled wristband applications vary when it comes to defense. These wristbands can be used to track wounded in warfare, track soldiers on battlefields, help families track relatives and identify victims during natural disasters. By employing RFID wristbands and satellite communication, medical personnel could better plan for arriving casualties. Some companies have been employing RFID wristbands to help families prevent their loved ones, who are at risk, from wandering away from the house. If an individual wearing the wristband approaches a door, an alarm sounds. This application is perfect for individuals suffering from Alzheimer's or similar conditions. RFID wristbands are becoming the standard in people identification and tracking. RFID enabled wristbands are used in healthcare, defense, entertainment industries and even some schools are starting to employ RFID to track student attendance.

10. RFID in Document Tracking

The documents are tracked as they leave the storage room and oftentimes within the storage room itself. And they're also being tracked as they pass through different locations of the building whether that's rooms or floors — wherever it's practical to be able to do that. The papers end up on person's desks; they end up in different filing cabinets, so it's very important to know exactly where each folder is at any given time because any person will need access to that folder at any given time. Tracking becomes very important in the locations where the folders are most commonly residing. People have significant traction in the hospital space where, although hospitals are moving to electronic documents, there's still a significant amount of hospitals and doctor's offices that are utilizing paper documents. There is also linking document tracking to that within the central IT management system. There's two standard frequencies that people use: either HF [high frequency], which is 13.56[MHz] with [ISO/IEC] 15693 protocol or UHF. People have seen both and there are definitely benefits to each one and, really, they're somewhat interchangeable within the user's facility. Within UHF what users can do is track large rooms so if they have an office they can track the files within a specific space so they know that the files are still residing in the specific space. The benefit of HF is localization. Users are able to localize exactly where that file is.

The museums and libraries need some specialized equipment like long-life chips with special adhesives that don't harm the documents or the archives. But there's definitely significant value in RFID tracking of those documents because, right now, the use case is that sometimes those documents are being lent to customers but they're not allowed to take them off of premises. The Label for archive, utilizes special glue that few vendors have certified through the ISO standard for archive documents.

11. RFID in pharmaceutical manufacturing

Since the U.S. Food and Drug Administration (FDA) issued a recommendation that pharmaceutical companies look at RFID and the use of e-pedigrees to help prevent counterfeiting of drugs, this issue has been high on the agenda of the industry. In 2005, a pilot program was carried out in the U.K. by Aegate, a pharmaceutical authentication service provider, in association with BT and DHL. Six U.K. pharmaceutical manufacturers—including Merck Generics UK, Merck Pharmaceuticals, Novartis, Schering Health Care, and Solvay—participated in the pilot, which examined using RFID and bar coding at the point of dispensing in 50 outlets as a way to detect fraudulent and counterfeit medicines. This pilot confined itself to authenticity of products from manufacture to dispensing to patients. A number of applications providers have launched solutions to enable pharmaceutical companies to track and trace products through the use of electronic pedigrees at the unit level. These developments may very well serve to accelerate the adoption of RFID tagging at the item level in pharmaceuticals, as both the means and the justification for adopting this strategy converge this year.

12. For industries that deal with perishable commodities

RFID is of high interest to the perishable goods industry because there are such tight deadlines in which to get goods to shelves in the retail operation. Some food products can spoil in as little as seven days. RFID tags can be used to track food products through the supply chain and gather data that can reveal how long it took products to move from one point to another. Ambient data, such as temperature or humidity, can also be captured to indicate how much shelf life is left for given products. Some in the industry are hoping to use this data to calculate when to replenish perishables and how much product to put on shelves. The benefits of applying RFID to perishable goods include improved food safety, more efficient product recalls, reduced costs due to less spoilage, lower inventories, more efficient logistics, and improved customer service.

***Using RFID in the blood and plasma market*:** 13.56 MHz RFID has been tested for blood and plasma applications with very good results. People are starting to look at UHF RFID as well: Liquids in general pose more challenges in UHF applications, however.

13. Various other RFID Applications

RFID Technology provides high value applications that solve problems in Security and Authentication. Whether the RFID application is Asset Tracking, such as a financial data center; Secure Credentials, such as the US Passport Card; Personnel Tracking, such as emergency evacuations and restricted zone access; or Vehicle Tracking, technology can provide the right tag for application.

***UHF Prelam for Passports cards*:** Few vendors designed and manufacture a product that meets the US Department of State's exacting RF performance requirements and the demanding card durability required for this 10-year life product.

Tamper Evident Container Sealing : For brand protection, loss prevention and homeland security, it is important to know that shipping containers have not been opened and resealed. RFID seal will make it immediately apparent if a container has been opened. This cargo seal has a long read range, low cost and a highly reliable tamper evident feature.

Healthcare - Labels for Tissue Trays : Small footprint, high read reliability and demanding printing and encoding were the design requirements for this label. Few vendors designed, built and converted this HF label which ensures that tissue samples are properly identified throughout the analysis and reporting process.

Worldwide UHF Asset Tags : The UHF frequency bands used in the US, the EU and Japan are not the same. This represents a major problem for manufacturers shipping products to multiple continents, since competitive RFID tags will not work in more than one market. Few vendors have solved this problem by developing a 'worldwide' tag that provides good read characteristics across the full UHF band, 860-960MHz.

Electronic Hologram : This software application provides authentication verification of tags in locations without being connected to a database. This provides brand owners the ability to monitor the integrity of their supply chains without integrating their data systems with those of their manufacturers, distributors and retailers. This software works within existing RFID standards; on existing RFID readers and chip hardware.

14. Some weird and funny applications

Preventing Toilets From Overflowing : What home would be complete without a high-tech toilet? Now user can purchase a "smart" toilet, one that shuts itself off when it is close to overflowing. According to AquaOne, their RFID-enhanced toilets are not only convenient, they prevent health risks in public places like hospitals and nursing homes. While the latter may be a bit of a stretch (who's licking the floor?), a toilet that never overflows sounds very appealing.

Identifying Human Remains : RFID tags are now being used to locate bodies in an effort to reunite loved ones. A morbid use for RFID, yes, but many people is appreciative when they are finally able to lay their friends and family members to rest.

Getting Into Nightclubs : There was a time when user merely had to look good or know someone special, but soon he might also need RFID tag implants to party in the VIP section. For example, the Barcelona hotspot Baja Beach Club is now grafting said tags into the arms of patrons who want instant access to the exclusive hangout. Not only will the tag identify user, it also functions as a debit card.

Cooking With Robots : Robotic pots and pans with RFID chips in the handles make are almost impossible to botch a meal. With these RFID chips, the cookware can be coordinated with a recipe card that has a similar chip. Then, the cookware will set its temperature and duration to the exact specifications the food.

Finding Lost Golf Balls : RFID application means that user will be able to pinpoint his ball's location; he still won't want to wade through a pond for it. However, long hikes through the rough will be a thing of the past.

Tracking Razor Blades : Due to low-cost RFID tags, Gillette can now afford to place small transponders in each package of their popular razor products. This is done in an attempt to salvage more razors as they make their way through a very convoluted supply chain. As the products are so small, many of them are lost or stolen.

Issuing Passports : To many people, this may be one of the scariest applications for RFID tags. The US State Department approved of passports with microchips inside and the technology is already being tested in trials.

With RFID becoming both smarter and smaller, the possibilities for its uses are endless. While this poses certain ethical dilemmas for government and commercial operations, it also simplifies life for the common person.

RFID COMPETING TECHNOLOGIES

RFID evolved as an attempt to provide a better alternative to bar code labels used extensively in products tracking. RFID thus has many inherent advantages over barcodes. Radio contact as required in RFID does away with the need of direct line of sight as required in bar codes. Moreover, with the capabilities of both read/write over the conventional read only, and the ability to store data and place RFID tags inside packing or embedding within products provides usage of RFID with many unique advantages. Given these advantages the question may arise why then RFID tags are not ubiquitous despite Wal Mart~Rs mandate that its top 100 suppliers should put radio frequency identification (RFID) tags on all pallets and cases they ship to its distribution centers and stores.

11
Near Field Communications

SO NEAR!

There are growing, numbers of applications where a form of very short range wire-less communications is needed. One technology that can meet this need is called Near Field Communications or NFC. It is able to provide contactless communication over short ranges, typically up to about 4 or 5 centimetres. In this way, it is able to provide connection simply by placing the two devices requiring connection close together. No physical connectors are used with NFC near field communication. The connection is more reliable and does not suffer problems of contact wear, corrosion and dirt experienced by systems caused by physical connectors. This wireless communication technology enables the exchange of data between devices over about a 10 centimeters (3.9 in) distance. The technology is a simple extension of the ISO/ IEC 14443 proximity-card standard (proximity card, RFID) that combines the interface of a smartcard and a reader into a single device. An NFC device can communicate with both existing ISO/IEC 14443 smartcards and readers, as well as with other NFC devices, and is thereby compatible with existing contactless infrastructure already in use for public transportation and payment. NFC is primarily aimed at usage in mobile phones.

Near Field Communications (NFC) is the contactless form of short range communication. It is being used in mobile phones, electronic devices and payment cards. NFC is basically a variation of other short-range wireless technologies already used throughout the world. Like RFID, near field communication can quickly swap information between devices when they're touched together. user can exchange things like text, images, URLs or other data simply by holding his phone up to various "smart tags."

EVOLUTION OF NFC TECHNOLOGY

NFC technology has evolved from a combination of contactless identification and interconnection technologies including RFID and it allows connectivity to be achieved very easily over distances of a few centimeters. Simply by bringing two electronic devices close together they are able to communicate and this greatly simplifies the issues of identification and security, making it far easier to exchange information. It is anticipated that NFC technology will allow the complex set-up procedures required for some longer range technologies to be avoided.

NFC is utilized in a whole variety of applications. These include:

* Personal computers
* Mobile phones, PDAs, etc
* Turnstiles
* ATMs
* Parking meters
* Check-out cash registers or "point-of-sale"

equipment

* Vending machines
* Applications around the office and house, e.g. garage doors, etc

USAGE IN MOBILE PHONES

NFC technology is currently aimed mainly at being used with mobile phones. There are currently three specific uses for NFC:

* Card emulation: the NFC device behaves like an existing contactless card
* Reader mode: the NFC device is active and reads a passive RFID tag, for example for interactive advertising
* P2P mode: two NFC devices are communicating together and exchanging information.

Various Mobile Phones applications are given as below :

* Mobile ticketing in public transport: an extension of the existing contactless infrastructure, such as Mobile Phone Boarding Pass.
* Mobile payment: the device acts as a debit/credit payment card.
* Smart poster: the mobile phone is used to read RFID tags on outdoor billboards.
* Bluetooth pairing: in the future pairing of Bluetooth 2.1 devices with NFC support will be as easy as bringing them close together and accepting the pairing. The process of activating Bluetooth on both sides, searching, waiting, pairing and authorization will be replaced by a simply bringing the mobile phones close to each other.

Future applications

* Electronic ticketing: airline tickets, concert/event

tickets, and others

* Electronic money
* Travel cards
* Identity documents
* Mobile commerce
* Electronic keys: replacements for physical car keys, house/office keys, hotel room keys, etc.
* NFC can be used to configure and initiate other wireless network connections such as Bluetooth, Wi-Fi or Ultra-wideband.

A patent licensing program for NFC is currently under development by Via Licensing Corporation, an independent subsidiary of Dolby Laboratories. A Public, platform independent NFC library is released under the free GNU Lesser General Public License by the name libnfc. Few years ago, in December 2008, the application eCL0WN was released which allows to read and copy the chip content of biometric passports. At the end of 2010 Telefónica, La Caixa and Visa participated in a pilot project in Sitges pioneer in Europe.

A further application that was proposed was that NFC connections could be used to configure the connection between two wireless devices. All that was required to configure them to operate together wirelessly would be to bring them together to effect the NFC "connection". This would initiate a set-up procedure; communication could take place over the NFC interface to configure the longer range wireless device such as Bluetooth, 802.11 or other relevant standard. Once, set up the two devices could operate over the longer range allowed by the second communication system. NFC is ideally placed to provide a link with the contactless smart card technology. It is broadly compatible with the existing standards that have been set in place. Accordingly it is quite possible that NFC enabled devices could be used for these applications as well. There are many other applications for NFC. These could include general downloading data from digital cameras or mobile phones.

Essential specifications

1. There are two modes
   * Passive Communication Mode: The Initiator device provides a carrier field and the target device answers by modulating the existing field. In this mode, the Target device may draw its operating power from the Initiator-provided electromagnetic field. The Target device functions as transponder.
   * Active Communication Mode: Both Initiator and Target device communicate by alternately generating their own fields. A device deactivates its RF field while it is waiting for data. In this mode, both devices typically have power supplies.
2. It requires working distance with compact standard antennas: up to 20 cm
3. Supported data rates: 106, 212, 424 or 848 kbit/s
4. As in ISO/IEC 14443, the NFC communicates via magnetic field induction, where two loop antennas are located within each other's near field, effectively forming an air-core transformer. It operates within the unlicensed radio frequency ISM band of 13.56 MHz. Most of the RF energy is concentrated in the allowed 14 kHz bandwidth range, but the full spectral envelope may be as wide as 1.8 MHz when using ASK modulation.
5. NFC devices are able to receive and transmit data at the same time. Thus, they need to check the radio frequency field and can detect a collision if the received signal does not match with the transmitted signal.

6. NFC employs two different codings to transfer data. If an active device transfers data at 106 kbit/s, a modified Miller coding with 100% modulation is used.

BEGINNINGS AND NFC FORUM

NFC technology has many of its roots in the RFID business. Some of the basic ideas came directly from RFID work that had been previously undertaken. It follows on from the proprietary smart card protocols and can be seen as an initiative to move forward the contact-less ticketing and payment applications. A couple of years ago, the standard for the technology was approved as an ISO/IEC standard on December 8 2003, which was approved earlier as an ECMA standard. The next stage in the standardisation process came when Nokia, Sony, and Phillips formed the NFC forum on 18th March 2004. The NFC forum grew quickly and in 2008 there were over 150 members comprising manufacturers, applications developers, financial services institutions, etc. During 2006, the NFC Forum formally outlined the architecture for NFC technology and released the first four Forum-approved specifications. These NFC specifications set in place a "road map" to enable interested parties to create their own products.

NFC BASICS

NFC technology provides short range wireless connectivity and has developed out of the RFID style contactless identification and interconnection technologies. The concept of NFC is that the technology is able to provide a very easy method of providing connectivity between devices that are a few centimetres apart, and without the need for all the security and set-up procedures that are needed for many other communication technologies. NFC is a wire-less technology, using radio frequencies. It operates at a frequency of 13.56 MHz within the globally available and unregulated 13.56 MHz frequency band. As a result no licenses are required for operation on these frequencies. In addition to this the radio transmissions using NFC are half duplex as the same channel is used for both transmit and receive. Also to prevent two devices transmitting together, they operate what is termed a listen before talk (LBT) protocol. The devices may only transmit if they previously listen to check that no other devices are transmitting. In view of the short ranges involved, the protocol used by NFC need not be as comprehensive as that used for together wireless protocols. Connection is made between two NFC devices when they are brought together so there is no difficulty in associating two devices. This occurs when the two devices are brought to less than about 4 centimetres of one another. The distances depend upon a variety of factors. The limit of 20 centimetres for the maximum communications distance has been realized. In this way a simple wave or touch can establish an NFC connection. NFC-enabled transactions are inherently secure as transmission range is so short.

To provide the standard interfaces, the underlying layers of NFC technology, it is necessary to follow the normal ISO standards. The data transfer rate may be either 106, 212 or 424 kbps. The initial communication speed may be changed later dependent upon the communication environment and the requirements. The NFC standard defines two types of NFC device. These are known as the Initiator and Target of the communication. As the names imply, the initiator is the device that initiates the communication and it controls the data exchanges. The Target device is the one that responds to the requests from the Initiator. The NFC standard defines two different modes of operation:

* **Active mode of communication:** In the active NFC mode of communication, both devices generate an RF signal on which the data is carried.

* **Passive mode of communication:** In this mode of communication, only one NFC device generates an RF field. The second passive device which is the target uses a technique called load modulation to transfer the data back to the primary device or initiator.

In addition to the NFC modes of operation, three communication modes are also defined:

* **Read / Write:** This mode of operation within NFC near field communication allows applications to transfer data in an NFC Forum-defined message format. It should be noted that this mode is not secure. It is also necessary to note that this mode is supported by the Contactless Communication API

* **NFC card emulation:** This NFC mode enables the NFC device to behave as a standard Smartcard. In this mode, data transfer is secure and the mode is also supported by the Contactless Communication API.

* **Peer to peer:** A third mode within NFC is the peer to peer mode which supports device to device link-level communication. It is worth noting that this mode of NFC communication is not supported by the Contactless Communication API.

Near-field Communication or NFC is a standard defined by the NFC Forum, a global consortium of hardware, software/application, credit card companies, banking, network-providers, and others who are interested in the advancement and standardization of this promising technology. NFC is a short-range radio technology that operates on the 13.56 MHz frequency, with data transfers of up to 424 kilobits per second. NFC communication is triggered when two NFC-compatible devices are brought within close proximity, around four centimeters. Because the transmission range is so short, NFC-based transactions are inherently secure; more on this shortly.

How do NFC compares to the other short-range communication technologies? The following table provides a comparison:

	NFC	RFID	IrDa	Bluetooth
Set –up time	<0.1ms	<0.1ms	~0.5s	~6 sec
Range	Up to 10cm	Up to 3m	Up to 5m	Up to 30m
Usability	Human centric Easy, intuitive, fast	Item centric Easy	Data centric Easy	Data centric Medium
Selectivity	High, given, security	Partly given	Line of sight	Who are you?
Use cases	Pay, get access, share, initiate service, easy set up	Item tracking	Control & exchange data	Network for data exchange, headset
Consumer experience	Touch, wave, simply connect	Get information	Easy	Configuration needed

Fig. 1: Comparing NFC to other close range communication technologies (Source: NFC Forum)

When compared to the other short-range radio technologies, NFC is extremely short ranged and *people-centric*. Some of the other short-range communication technologies have similar characteristics, for example RFID, while others are completely different yet complimentary to NFC; for example Bluetooth and Infrared. A good scenario of such compliment is the combination of NFC and Bluetooth, where NFC is used for pairing (authenticating) a Bluetooth session used for the transfer of data.

STANDARDS AND CAPABILITY

With Near Field Communications set to become widely accepted in many applications, the system has been standardised by a number of globally accepted standards bodies. NFC has standards accepted by ISO (18092), ECMA (340) and ETSI. Additionally NFC is compatible with Philips' MIFARE (ISO 14443 A) and Sony's FeliCa smart card protocols. Although Near Field Communications, NFC, may at first sight appear to be, "just another wire-less system", or similar to RFID this is not the case and it appears that it will find a number of applications. Applications are now becoming more widespread and further applications are following on.

NFC STANDARDS

NFC is a standard, and is ISO standards-based. The ISO 14443 Type A and Type B standards + FeliCa is a four-part international standard for contact-less smart cards operating at 13.56 MHz in close proximity with a reader antenna. The ISO 18092 standard defines communication modes for NFC Interface and Protocol. NFC was approved as an ISO/IEC standard on December 8, 2003 and later as an ECMA standard. NFC is an open platform technology standardized in ECMA-340 and ISO/IEC 18092. These standards specify the modulation schemes, coding, transfer speeds and frame format of the RF interface of NFC devices, as well as initialization schemes and conditions required for data collision-control during initialization for both passive and active NFC modes. Furthermore, they also define the transport protocol, including protocol activation and data-exchange methods. The air interface for NFC is standardized in:

* ISO/IEC 18092 / ECMA-340
  — *Near Field Communication Interface and Protocol-1* (NFCIP-1)
* ISO/IEC 21481 / ECMA-352
  — *Near Field Communication Interface and Protocol-2* (NFCIP-2)

NFC incorporates a variety of existing standards including ISO/IEC 14443 both Type A (normal) and Type B (banking/short range), and FeliCa. NFC enabled phones work basically, at least, with existing readers. Especially in "card emulation mode" a NFC device should transmit, at a minimum, a unique ID number to an existing reader. In addition, the NFC Forum has defined a common data format called NFC Data Exchange Format (NDEF), which can store and transport various kinds of items, ranging from any MIME-typed object to ultra-short RTD-documents, such as URLs. NDEF is conceptually very similar to MIME. It is a dense binary format of so-called "records", in which each record can hold a different type of object. By convention, the type of the first record defines the context of the entire message.

Few Initiations

GSMA: The GSM Association (GSMA) is the global trade association representing 700 mobile phone operators across 218 countries of the world. They have launched 3 initiatives.

The Mobile NFC initiative: fourteen mobile network operators, who together represent 40% of the global mobile market, back NFC and are working together to develop NFC applications. They are Bouygues Télécom, China Mobile, AT&T, KPN, Mobilkom Austria, Orange, SFR, SK Telecom, Telefonica Móviles España, Telenor, TeliaSonera, Telecom Italia Mobile (TIM), Vodafone, Deutsche Telecom.

StoLPaN: StoLPaN ('Store Logistics and Payment with NFC') is a pan-European consortium supported by the European Commission's Information Society Technologies program. StoLPaN will examine the as yet untapped potential for bringing together the new kind of local wireless interface, NFC and mobile communication.

Other standardization bodies that are involved in NFC include:

* ETSI / SCP (Smart Card Platform) to specify the interface between the SIM card and the NFC chipset.
* GlobalPlatform to specify a multi-application architecture of the secure element.
* EMVCo for the impacts on the EMV payment applications.

COMPARISON WITH BLUETOOTH

The following table provides a comparison between NFC and Bluetooth:

	NFC	Bluetooth	Bluetooth Low Energy
RFID compatible	ISO 18000-3	active	active
Standardisation body	ISO/IEC	Bluetooth SIG	Bluetooth SIG
Network Standard	ISO 13157 etc.	IEEE 802.15.1	IEEE 802.15.1
Network Type	Point-to-point	WPAN	WPAN
Cryptography	not with RFID	available	available
Range	< 0.2 m	~10 m (class 2)	~1 m (class 3)
Frequency	13.56 MHz	2.4-2.5 GHz	2.4-2.5 GHz
Bit rate	424 kbit/s	2.1 Mbit/s	~1.0 Mbit/s
Set-up time	< 0.1 s	< 6 s	< 1 s
Power consumption	< 15mA (read)	varies with class	< 15 mA (xmit)

As the table reflects that NFC and Bluetooth are both short-range communication technologies which are integrated into mobile phones. NFC can be used for the set-up of wireless technologies, such as Bluetooth. NFC sets up faster than standard Bluetooth, but is not much faster than Bluetooth low energy. Instead of performing manual configurations to identify devices, the connection between two NFC devices is automatically established quickly. It happens in less than a tenth of a second. The maximum data transfer rate of NFC (424 kbit/s) is slower than that of Bluetooth V2.1 (2.1 Mbit/s). Shorter range reduces the likelihood of unwanted interception. NFC is suitable for crowded areas where correlating a signal with its transmitting physical device (and by extension, its user) becomes difficult. In contrast to Bluetooth, NFC is compatible with existing passive RFID (13.56 MHz ISO/IEC 18000-3) infrastructures. NFC requires comparatively low power, similar to the Bluetooth V4.0 low energy protocol. However, when NFC works with an unpowered device (e.g. on a phone that may be turned off, a contactless smart credit card, a smart poster, etc.), the NFC power consumption is greater than that of Bluetooth V4.0 Low Energy.

THE CONTACTLESS COMMUNICATION API

The Contactless Communication API Java specification, led by Nokia which is defined under the Java Community Process as JSR-257, defines a set of APIs for proximity, contactless-based communication. The API consists of five Java packages, as follows:

JSR 257 Java Packages

It should be noted that the only mandatory package is javax.microedition.contactless. The Contactless Communication API allows to Discover and Exchange data with contactless targets such as NDEF tags, RFID tags, and external smartcards. The API also provides support for visual tags. The comparison and description of internal and external elements is as following:

Table - External Readers and Secure Elements

External Readers	Secure Elements
External readers include contactless payment readers in Point of Sale stations, ticketing systems on transportation systems, external radio, visual tags such as NFC, RFID and barcodes, or Smartcards	Secure elements (SE) can be internal or external elements; example of a secure element is a Java Card-based smartcard. MIDlets can access secure elements by using the Security and Trust Services API (SATSA), and/or the Contactless Communication API (JSR 257). External readers access internal secure elements directly via the RFID circuitry (using the Card Emulation mode).

(*Source:* JSR-257 Specification)

Table - JSR 257 Java Packages

Java Package	Interfaces	Classes	Exceptions
javax.microedition.contactless A **mandatory** package that contains all the target discovery and classes common to all targets	TagConnection TargetListener TargetProperties TransactionListener	DiscoveryManager TargetType	Contactless Exception
javax.microedition.contactless.ndef An optional package for communicating with NDEF formatted data tags	NDEFRecordListener NDEFTagConnection	NDEFMessage NDEFRecord NDEFRecordType	
javax.microedition.contactless.rf An optional package for communicating with RFID (no NDEF formatted data) tags	PlainTagConnection		
javax.microedition.contactless.sc An optional package for communicating with external smartcards	ISO14443Connection		
javax.microedition.contactless. visual An optional package for reading and generating visual tags	ImageProperties VisualTagConnection	SymbologyManager	VisualTag CodingException

(Source: JSR-257 Specification)

Using the Contactless Communication API

The Contactless Communication API allows *discovering* and *exchanging data* with supported contactless radio and visual targets (tags). The contactless communication application takes the following course in general:

1. The application queries the implementation to discover the target types that are supported by the handset, For each supported target type, the application can register a target listener to receive activity notifications. Alternatively, the application can register with the *PushRegistry* for activation due to target activity; supported target activities are NDEF and secure element in card emulation mode activities,
2. As targets come into proximity, they are detected (discovered) by the implementation, which in turn notifies the application(s) by invoking the appropriate activity listeners. Alternatively the PushRegistry activates the MIDlet, now for each discovered target, the application can learn the target's properties, the application can connect to, read, write and exchange data with the discovered target,
3. When done, to release resources, the application closes any opened connections.

The Contactless Communication API and the Generic Connection Framework

The Contactless Communication API defines the following connection-related interfaces:

1. TagConnection defines the base interface for all RFID, smartcards, and NFC connections:
   a. ISO14443Connection extends TagConnection and is used for contactless connection with ISO 14443-4 compliant contactless smart card

   b. PlainTagConnection extends TagConnection and is used for contactless connection with RFID (non-NFC) tags
   c. NDEFTagConnection extends TagConnection and used for contactless connection with NFC Forum formatted RFID tags and contactless smart cards.
2. VisualConnection is used for contactless connection to a visual tag such as a barcode

Discovering and Listening for Supported Targets

To discover targets that are supported by the platform invoke the following method:

discoveryManager.getSupportedTargetTypes(),

This returns an array of TargetTypes. Then for each target type of interest, it registers a target listener.

Using PushRegistry to Launch NFC Applications

Automatic application activation or startup is a very important aspect for the success and acceptance of NFC-based applications. It allows for a better user experience. By just coming into proximity of (or touching) a reader or another NFC device, the application is automatically started. The PushRegistry facility that is part of the Mobile Information Device Profile (MIDP) provides for the automatic application-launch based on timers and connection activity. The Contactless Communication API extends the PushRegistry for application startup only for 1) NDEF record types and 2) for secure element (activity) card emulation mode communication. To ensure that the application when launched receives notifications related to discovered NDEF records or card emulation activity, it must register its NDEFRecordListener and TransactionListener as soon as it is started.

Security Considerations

The Contactless Communication API is subject to the security policies of the underlying mobile platform. The user request permissions via the JAD file or the JAR manifest, by creating MIDlet-Permissions property entries, using the permission names in the table above. Attempting to use a restricted operation without proper permission causes the system to throw a SecurityException. For implementations of the Contactless Communication API on CDC profiles, the application must use java.security.Permission-based security checks. The following methods require permissions:

Security aspects : Although the communication range of NFC is limited to a few centimeters, NFC alone does not ensure secure communications. NFC offers no protection against eavesdropping and is also vulnerable to data modifications. Applications have to use higher-layer cryptographic protocols (e.g., SSL) to establish a secure channel.

Understanding Eavesdropping : Eavesdropping is extremely affected by the communication mode. A passive device that does not generate its own RF field is much harder to eavesdrop on than an active device. One Open source device that is able to eavesdrop on passive and active NFC communications is the Proxmark instrument. The RF signal for the wireless data transfer can be picked up with antennas. The distance from which an attacker is able to eavesdrop the RF signal depends on numerous parameters, but is typically a small number of meters.

Relay attack : Because NFC devices usually include ISO/IEC 14443 protocols, the relay attacks described are also feasible on NFC For this attack the user has to forward the request of the reader to the victim and relay back its answer to the reader in real time, in order to carry out a task pretending to be the owner of the victim's smart card. One of libnfc code examples demonstrates a relay attack using only two stock commercial NFC devices. Once lawfully opened, access to a secure function or data is usually protected by time-out closing after a period of inactivity. Modern attacks may succeed, despite provisions to shut down access when the user turns inactive. The distance of a successful attacker to the locus of lawfully granted access is not addressed with any of the described concepts.

NFC PHONE APPLICATION

As it is explained that NFC is a short-range wireless connectivity technology standard designed for intuitive, simple and safe communication between electronic devices. NFC communication is enabled by bringing two NFC compatible devices within a few centimeters of one another or for the two devices to literally "touch" one another. Applications of NFC technology include calendar synchronization or electronic business cards and access to online digital content. NFC technology evolved from a combination of contactless identification and interconnection technologies. It combines the functions of a contactless reader, a contactless card and peer-to-peer communication. Near Field Communication (NFC) is a standards-based, short-range (a few centimeters) wireless connectivity technology that enables simple and safe two-way interactions among electronic devices, allowing consumers to perform contactless transactions, access digital content and connect electronic devices with a single touch.

Based on inductive-coupling (such as Powermat and other wireless charging schemes), NFC uses loosely coupled inductive circuits to exchange power and/or data over a short distance (usually about four centimeters). While it shares the same basic technology with things like RFID tags and contactless smartcards, NFC is implemented as a read-only tech in the Nexus S. That means user will only get information *from* other NFC tags, not the other way around. There are literally hundreds of use cases for NFC.

NFC Phones using single wire Protocol and UICC

There are already several phones out in the market using the UICC as the secure element today (08/2008): The Sagem my700X, the LG L600V, the Nokia 6131 SWP and the Motorola SLVR L7. All devices are more or less concept devices and come with an Inside NFC Chip inside. The device had been shown already at Cartes 2007 and Mobile World Congress 2008. In order to develop applications with these devices an SDK (like the Gemalto Developer Suite) and a SWP UICC is required. All four devices are already capable of using SCWS. The Motorola, the Sagem as well as the LG are used in the Payez Mobile NFC Trial in Strassbourg and Caen. The Sagem was used in a Trial with the metro in Paris too. The Touch & Travel Trial from Deutsche Bahn uses the Motorola.

APPLICATIONS IN DAILY LIFE

In the realm of new technologies, NFC has clear potential and practical uses. Nokia, Sony, and Royal Philips Electronics founded the NFC Forum in 2004 in order to promote the short-range wireless connectivity technology. Samsung, Motorola, Microsoft and more than 140 other organizations all joined the party shortly after. NFC allows a device, usually a mobile phone, to collect data from another device or NFC tag at close range. In many ways, it's like a contactless payment card that is integrated into a phone. In other ways, it's similar to Bluetooth, except that instead of programming two devices to work together, they can simply touch to establish a connection.

1. Contactless Payment

Unlike many other wireless technologies, NFC has a short range of about 1.5 inches. This makes it a good choice for secure transactions, such as contactless credit card payments. MasterCard and Visa are both members of the NFC Forum, and both companies have been involved in pilot programs that use NFC-enabled phones as a flash payment option. The "digital wallet" concept could extend to coupons and other offers. Consumers can use the site to download coupons, which they exchange by having their phone swiped at the point of purchase. The retailers benefit from being able to track who their coupons are sent to and how they are used.

2. Transportation

NFC works with most contactless smart cards and readers, meaning it could easily be integrated into the public transit payment systems in cities that already use a smart card swipe. In 2008, German rail operator Deutsche Bahn launched an NFC-ticketing pilot program in which 200 travelers touched their phones to an NFC tag when they boarded the train and then to another when they got off. The fare was calculated and added to their monthly bill. In January 2010, the successful program was expanded to an additional 3,000 travelers. Madrid planned to start a similar pilot program with its bus system in 2010.

3. Health Care

Not only can NFC tags provide medical professionals with information about what treatments a patient should receive, but they can also keep track of when nurses and doctors have checked in with that patient and when. Each time the tag is scanned, the information about who scanned it and when can be transferred to a database. In addition to improving treatment, NFC tags also have potential in the research realm.

4. Smart Objects

An NFC tag often contains information like a phone number or URL. One of the largest series of experiments that uses phones to pick up information from tagged locations is SmartTouch, a project funded under the European ITEA research program between 2006 and 2008. Most of the trials took place in Oulu, Finland, where the city installed about 1,500 "infotags" that could be read with a mobile phone. For another project, infotags were installed in schools. Students could get their individual daily schedule, announcements, and information about homework by waving their phones past the tags. A trial held in one pub allowed customers to tap cards with their NFC-enabled phones for more information about products. NFC may have similar applications as bar codes do now. But being able to add more information to any object by integrating a tag has led to some interesting applications that go far beyond billboards.

5. Social Media

Before Foursquare took off, a German company called Servtag was working towards a similar concept for NFC-enabled phones called Friendticker. The company applied more than 250 NFC-tag stickers at various locations in Berlin that users would swipe their phones past in order to alert their friends that they were "checked in" at that location. While Foursquare may have stolen the thunder for location-based networking, there are still plenty of social media applications for NFC in the works. Last year, a German university (Technische Universität München) submitted a prototype to the NFC Forum competition that integrated with Facebook. The application, NFriendConnector, allowed people who met in a physical space to exchange profile data through their phones. Instead of stalking a new acquaintance's profile after a night out, this application provides an option to run a matching method based on variables the user provides while still chatting with them in the bar.

12

RFID in Indian and Global Perspectives

USAGE AND CONCERNS!

The successful usage of the technology over a considerable period of time was an important factor in several parts of the world in determining its reliability. The passive RFID technology (widely used across the globe) may face certain unexpected challenges in India. With its economy growing like never before it is offering immense opportunities for everybody so how could RFID stay behind. The need for retail and manufacturing companies is to better track inventory and raw materials. The decreasing costs of microchips have led to enhanced interest in automatic identification systems. Retailers continue to face challenges in managing their business due to inconsistencies in data. In addition, retailers aren't always able to quantify key performance indicators across the enterprise. They have difficulty providing including sales, marketing and operational metrics across the enterprise.

Indian perspective

Slowly and steadily RFID is penetrating India and from defense sector to corporate and even farmers are making uses of it. Automatic identification technologies try to overcome this by allowing machines to identify objects and capture information, providing organizations with much greater degrees of control and flexibility in managing goods as they move through the supply chain. Even the government of India is looking for RFID deployment and will be conducting RFID tests for cold chain management. Corporates have been testing RFID in a number of areas such as library management, asset tracking for computers to document management. Besides frequency related problems, India's varying and extreme climatic conditions may also prove to be a major hurdle in adoption of RFID technology for electronic toll collection.

Recently the Ministry of Road Transport and Highways has decided to adopt the passive RFID (Radio Frequency Identification) technology for electronic toll collection on national highways. The adoption of the technology, adhering to Gen 2, ISO 18000 6C standards, is expected to ensure seamless travel across the country and also plug leakages in the toll collection system. The passive RFID system has two components-tags and readers. The tamper proof passive tags are pasted on the windscreen of vehicles and the readers are installed at toll plazas. In India, the cost of each passive tag would be Rs 100 initially. This could come down to Rs 10 over time with increase in volume. In the future, passive RFID tags may be installed in all vehicles during the manufacturing stage. The cost of each reader, on the other hand, would be around Rs 2 lakh. Each lane at a toll plaza would have to be fitted with at least one reader. Some toll plazas may opt for two readers per lane—one for checking the authenticity of the tag on the vehicle and the other for reading information to debit charges from the vehicle's account. In India, mobile phone operators use both GSM and CDMA technologies. The GSM technology works in the frequency bands of 900 MHz and 1800 MHz and CDMA works in the frequency band of 800 MHz. The RFID technology will use the 860 MHz-960 MHz range. It is therefore likely to cause interference with mobile phone operations. In the US and other countries in Europe, where the RFID technology is being successfully used, this problem does not exist. The RFID tags are generally manufactured in the US and in European countries. The tags may have to be modified in accordance with climatic conditions in India. It has to be kept in mind that many parts of the country experience extreme heat and humidity during summer months. The tags must be able to withstand the extreme climatic conditions. The Ministry of Road Transport and Highways plans to introduce the passive RFID technology for electronic toll collection by May 2012. Mahindra & Mahindra, an automobile company in the India was one of the early adopters of RFID in areas ranging from warehousing to vendor tracking. The booming retail sector in the country is too eyeing RFID implementation. One of the reasons for the adoption of the technology in the country is the falling costs and secondly the semiconductor policy of the government is favoring it.

One of the most popular applications of RFID is in tracking assets. A wide range of RFID system is used for an effective asset management programme. In India, very soon RFID intervention will be used extensively. Neighbourhood Post offices using RFID to track lost parcels, Indian farmers keeping a tab on their cattle using RFID, these are two small examples of massive RFID deployment. Other applications include legal firms turning to RFID for tracking or automating files, airlines use the technology to track the location of luggage trolleys, construction companies using RFID tags inserted to concrete blocks for instant dissemination of results over the internet. Department of Information Technology has its focus on the R&D areas of RFID and providing RFID based techno solutions to the Indian industries. With this objective, the ambitious "National RFID Program" project was initiated in April, 2007. The program was being implemented jointly by IIT, Kanpur, C-DAC, Noida and SAMEER, Mumbai. Research component is mainly undertaken by IIT Kanpur, application development including deployment and support is provided by C-DAC, Noida while Mumbai is focused primarily on RFID antenna design and other RF related issues. C-DAC and IIT, Kanpur also undertook RFID related manpower development programs. RFID characterization/modifications includes characterization of RFID System, providing technical assistance in field trials, designing of Reader Antenna for Gen II Tags, Linear Polarized Planar Antennas and Circular Polarized Antennas and developing RFID System Using Zigbee Modules.

India's defense ministry is embarking upon a pilot program aimed at developing radio-frequency identification (RFID)-based smart cards for Indian army, navy and air force personnel. The ministry will choose a consultant contractor for the project, which will have to analyze existing access control systems, infrastructure, and processes across locations and software requirements for each military service. Ministry sources tell Aviation Week that the high-profile exercise is being initiated to tighten security across all three branches of the Indian military. The RFID cards will include the latest biometric features; an initial batch of 3,000 will be produced for three designated locations. The Indian Institute of Technology (Kanpur) will be involved with the development of smart card operating system (SCOSTA-CL). Based on the OS provided by IIT-Kanpur, an algorithm and design architecture will be formulated.

There is another hope of a new research collaboration established by the McMaster RFID

Applications Lab (MRAL), the Indian Institute of Technology Delhi (IIT- New Delhi), IPICO Inc. and Strategic Consultants (New Delhi). The $2.5 million initiative is receiving $1.2 million in matching funding by International Science and Technology Partnerships Canada (ISTPCanada) and the Global Innovation & Technology Alliance (GITA). Wipro Technologies launched an RFID-enabled retail store on its sprawling corporate campus in Bangalore, India. The concept store is designed to demonstrate how radio frequency can automate checkout, stock maintenance, and the tracing and tracking of goods in a clothing shop. There is not a tidbit of surprise in the announcement. Wipro seems to be aiming to be the RFID integration leader in India — and other parts of the world, too. RFID deployment in the country has also started. Pantaloon is piloting an RFID solution at its Tarapur warehouse and factory. Wipro Infotech is the implementation partner and has been involved in designing the architecture. The RFID system has been integrated with Pantaloon's existing infrastructure and the company expects the RFID solution to help it improve collaboration across the supply chain from the point of sale onwards.

With it's varied usage across sectors, RFID can be termed as one of the most promising and anticipated technologies in recent years poised for nothing less than complete transformation in the supply chain. EPCglobal India, a division of EAN India, is a regional member organisation of EPCglobal Inc, the not-for-profit body set up to establish global standards regarding the development, implementation and adoption of Electronic Product Code (EPC) and Radio Frequency Identification (RFID) technology, and support of the EPCglobal Network. EAN India is a registered society, promoted by the Ministry of Commerce, Government of India. It is a joint Government-Industry initiative at promoting use of international identification and communication standards and best practices within the industry, which can further enhance efficiency of their supply and demand chains through use of business-led EAN.UCC standards.

Retail shrinkage is the difference between book stock and actual stock. It is the unaccounted loss of retail goods. Its main causes are theft by employees, administrative errors, shoplifting by customers or vendor fraud. Large retail outlets such as Big Bazaar and Pantaloon have investments in RFID, CCTV and antennas to reduce retail shrinkage. This is where RFIDs are useful in protection of goods. India's organized retail is only 3 percent while 97 percent is unorganized. India is the second largest growing economy in retail, after China. New products launched globally are now launched simultaneously in India as well. However, RFID has its own share of defects. Some RFID tags cannot be detected by the antennas if they are shielded by the hand or the body. Another issue is threat to privacy. RFID can be used to trace customer behavior or find customer specific information. The tags can be read even if they are kept in the cars or homes of the customer. RFID is responsible for transforming the retail scenario in India from traditional to modern. RFID did not make a significant mark in jewelry segment—at least, not in India. The technology itself was not considered mature or reliable enough to handle expensive items like jewelry, for even a little flaw in the system could be catastrophically expensive to a jewelry firm. Despite these hurdles, the acceptance scenario for India's jewelry companies has also evolved principally because people have seen or heard about successful deployments of RFID within that industry, and about the associated positive return on investment (ROI).

To support the wishes and expectations of jewelry merchants, the technology has also evolved to a much higher level. Now we can expect anti-collision capabilities in high-frequency (HF) RFID, thus enabling interrogators to read multiple RFID tags simultaneously. Some RFID hardware vendors have enhanced the hardware to a level at which tags can be read even when stacked one on top of another. For instance, RFID can be useful in tracking the performance of salespeople, in terms of time and the actual business he or she gets. Real-time inventory management and the movement of stock can provide some crucial data that can very useful for production management, shelf management and stock management.

India represents an economic opportunity on a massive scale, both as a global base and as a domestic market. Retail growth in the coming years is expected to be stronger than GDP growth. In this light, Radio Frequency Identification (RFID) is fast becoming a hotbed of activity targeting the Indian market in retail, logistics and manufacturing. There is growing interest and investment in the use of radio frequency identification (RFID) technology as a means of greater visibility in supply chains, leading to improved inventory management through improved customer and sales information. Retailing is emerging as a technology-intensive industry, and China and India are emerging as potential major retail countries. Reliance Retail, one of India's largest retailers, has been testing RFID technology and is preparing applications for use at a large number of its hypermarkets and supermarkets, as well as its electronics and convenience stores. Reliance Retail has drawn up plans for using RFID to support its operations. The company has developed five RFID-deployment scenarios, including the tracking of reusable crates of fresh food; item- and case-level tracking of high-value goods; and pallet and case tagging of various goods. Reliance has performed proof-of-concept tests using passive EPC Gen 2 UHF RFID tags to track crates of fresh foods and cases of high-value goods. The company managed to achieve above 90 percent read rates and found that RFID helped it reduce shipping and receiving errors while increasing productivity.

Some of the expert observations of Marshall Kay, founder of Wilmette, Ill.-based consulting practice RFID

Sherpas and former director of the North American RFID program for Kurt Salmon Associates, are as follows

* "There is much more happening than would be apparent to a casual observer, and it has been this way for almost two years,"
* "Item-level RFID is drawing increasing interest from luxury retailers and big-box discounters alike. This reflects a growing recognition of the many ways that value is generated once an item is tagged."

Changes in retail RFID over the past few months are as following:

1. The number of retail CEOs who have taken a direct interest in RFID has grown.
2. Disposable RFID tags are being used in the United States on items that retail for under $5.
3. There are reports that auditors are now willing to waive the requirement of laborious physical inventory counts if retailers instead track their inventory throughout the year with RFID. Not only can retailers save the money they pay outside parties to perform these counts, they can also save the overtime wages they pay their own employees to shadow these people.

The fact that a technology can be helpful in multiple ways should not be held against it. One of the problems that RFID quickly solves — inaccuracy of retail perpetual inventory data — is a chronic problem that has plagued the retail industry for decades. Specialty retailers will also be reaping the rewards of inventory efficiency and visibility across their supply chains, further strengthening their balance sheets and making them even more dangerous competitors. We are yet to see the adoption of RFID products in the logistics and retail space on a large scale. The primary reason being, RFID currently competes with the low cost Barcode technology, which can be printed on just anything for ID purpose. However, using RFID involves the cost of the chip, affixing it to the inventory etc. Usage of RFID in retail is still not economically viable as yet, so we at ST are currently only focusing on RFID deployments at projects which already have a validated economic model. Some of these projects include tagging books in a library to maintain the inventory and segregate the books as required.

Global Perspectives

In 2010 three key factors drove a significant increase in RFID usage: decreased cost of equipment and tags, increased performance to a reliable 99.9% and a stable international standard around UHF passive. At RFID Journal Live 2010 in Orlando, Airbus detailed 16 active projects being conducted by ODIN technologies, IBM and - most recently added to the team - CSC. The two other areas of significant use are financial services for IT asset tracking and healthcare with more than 60% of the top medical device companies using passive UHF RFID in 2010. RFID is becoming increasingly prevalent as the price of the technology decreases. The Japanese HIBIKI initiative aims to reduce the price to 5 Yen (4 eurocents). In January 2009 Envego announced a 5.9 cent tag, and in March 2010 a Korean laboratory successfully created a printed chip using carbon nanotubes that would have the price of a passive UHF RFID tag to about three cents by late 2011.

Since summer 2009, two credit card companies have been working with Dallas, Texas, based DeviceFidelity to develop specialized microSD cards. When inserted into a mobile phone, the microSD card can be both a passive tag and an RFID reader. After inserting the microSD, a user's phone can be linked to bank accounts and used in mobile payment. Dairy Queen in conjunction with Vivotech has also begun using RFIDs on mobile phones as part of their new loyalty and rewards program. Patrons can ask to receive an RFID tag to place on their phone. After activation, the phone can receive promotions and coupons, which can be read by ViVOtech's specialized NFC devices.

Governments use RFID applications for traffic management, while automotive companies use various RFID tracking solutions for product management. Many of these solutions may work together in the future, though privacy regulations prevent many initiatives from moving forward at the same pace that technology allows. Following a successful pilot, Housing & Development Board (HDB) Singapore called two tenders in 2006 to implement RFID to replace the paper Season Parking Ticket (SPT). The successful tenderers have distributed RFID tags to SPT holders since March 2007. In Vietnam, Futech has an auto checking ticket system in place for many buildings in the country.

RFID is being used for E - Tolling in Motorways, Implemented by NADRA. In Dubai, UAE, RFID is being used for E - Tolling - SALIK in Motorways, Implemented by RTA. In Turkey, RFID has been used in the motorways and bridges as a payment system since Nov 2008; it is also used in public transportation systems in Çanakkale, Izmir and Denizli. RFID is used in Malaysia Expressways payment system. The name for the system is Touch 'n Go. As the system's name indicates, the card is designed to only function as an RFID card when the user touches it. In Norway, all public toll roads are equipped with an RFID payment system known as AutoPASS. In Italy, all public toll roads are equipped with an optional RFID payment system named as Telepass. In Ireland, the eToll system uses RFID tags for payments on all road tolls, including the barrier-free M50 toll between exits 6 and 7. In Singapore, public transportation buses and trains employ passive RFID cards known as EZ-Link cards. Traffic into crowded downtown areas is regulated by variable tolls imposed using an active tagging system combined with the use of stored-value cards (known as CashCards). In Ontario, Canada, Electronic Road Pricing systems are used to collect toll payments on Ontario Highway 407. Florida's SunPass, various systems in Texas including D/FW's NTTA TollTag, the Austin metro TxTag and Houston HCTRA EZ Tag (which as of early 2007 are all valid on any Texas toll road), Kansas's K-Tag, The "Cross-Israel Highway" (Highway 6), Philippines South Luzon Expressway E-Pass, Brisbane's Queensland Motorways GoVia tag (previously called E-Toll) System in Australia, Autopista del Sol (Sun's Highway), Autopista Central (Central Highway), and every forthcoming urban highway (in a "Free Flow" modality) concessioned to private investors in Chile, all toll tunnels in Hong Kong (Autotoll) and all highways in Portugal (Via Verde, the first system in the world to span the entire network of tolls), France (Liber-T system), Italy (Telepass), Spain (VIA-T), Brazil (Sem Parar - Via Fácil). The tags, which are usually the active type, are read remotely as vehicles pass through the booths, and tag information is used to debit the toll amount from a prepaid account. The system helps to speed traffic through toll plazas as it records the date, time, and billing data for the RFID vehicle tag.

Throughout Europe, and in particular in Paris (system started in 1995 by the RATP), Lyon, Bordeaux, Grenoble, Nancy and Marseilles in France, in the whole of the Portuguese highway system and in many Portuguese public car parks, Milan, Turin, Naples and Florence in Italy, and Brussels in Belgium, RFID passes conforming to the Calypso international standard are used for public transport systems. They are also used now in Canada (Montreal), Mexico, Israel, Bogotá and Pereira in Colombia, Stavanger in Norway, Luxembourg, Gävle in Sweden, etc. The card can be recharged with cash at add-value machines or in shops, and can be read several centimetres from the reader. The same applies for Delhi Metro, the rapid transit system in New Delhi, capital city of India. In Shanghai the Shanghai Public Transportation Card allows the user to credit money in advance and to be debited according to the distance travelled, as determined by the check-in and check-out stations. The card can also be used to pay taxi drivers, and some shops offer card readers as well.

The Moscow Metro, the world's second busiest, was the first system in Europe to introduce RFID smartcards in 1998. The Washington Metro rail became the first U.S. urban mass-transit system to use RFID technology when it introduced the SmarTrip card in 1999. JR East in Japan introduced SUICa (Super Urban Intelligent Card) for transport payment service in its railway transportation service in November 2001, using Sony's FeliCa (Felicity Card) technology.

In the United States, the Chicago Transit Authority has offered the Chicago Card and the Chicago Card Plus for rail payments across the entire system since 2002 and for bus payments since 2005. The MBTA introduced the RFID enabled CharlieCard across Boston's subway, streetcar, and bus system in 2006, replacing the decades-old token based fare collection system. The New York City Metropolitan Transportation Authority conducted an RFID trial that utilized PayPass by MasterCard. In the UK, operating systems for prepaying for unlimited public transport have been devised, making use of RFID technology. The design is embedded in a creditcard-like pass, that when scanned reveals details of whether the pass is valid, and for how long the pass will remain valid. In Canada, Metrobus in St. John's adopted RFID on December 1, 2006. In the Greater Toronto and Hamilton areas, under Metrolinx, full implementation of the RFID farecard Presto began in November 2009, and will be rolled out in stages across the network.[19] OC Transpo in Ottawa is also in the process of implementing Presto, with completion expected in late 2011.[20]

In Oslo, Norway, the upcoming public transport payment is to be entirely RFID-based. The system was slated for introduction around spring 2007. The Transperth public transport network in Perth, Western

Australia uses RFID technology its SmartRider ticketing system, allowing passengers to "tag on" and "tag off" and be charged automatically, according to how many zones they have travelled. In Santiago (Chile) the subway system Metro and the recently implemented public transportation system Transantiago use an RFID card called "Bip" or "Multivia". In Cali (Colombia) the recently-implemented card system for the Masivo Integrado de Occidente(MIO) system uses an RFID card. In Dublin (Ireland) the LUAS light rail system has been using an RFID enabled 'smart card' system since March 2005. In Seattle the Orca Card was introduced in 2009 for fares on buses, ferries, light rail, a street car, and commuter trains. In Tacoma, Washington, a sticker tag is used for paying the toll of the Tacoma Narrows Bridge.

Since 2010, bus transit in Ljubljana (Slovenia) is payable only by RFID with pre-paid city card named Urbana which can be re-filled with monthly passes or cash value on Urbanomats scattered all over the city. In Gothenburg (Sweden), public transportation has used RFID cards since 2006. RFID combined with mobile computing and Web technologies provide a way for organizations to identify and manage their assets. Mobile computers, with integrated RFID readers, can now deliver a complete set of tools that eliminate paperwork, give proof of identification and attendance. Web based management tools allow organizations to monitor their assets and make management decisions from anywhere in the world. Web based applications now mean that third parties, such as manufacturers and contractors can be granted access to update asset data, including for example, inspection history and transfer documentation online ensuring that the end user always has accurate, real-time data.

RFID use in product tracking applications begins with plant-based production processes, and then extends into post-sales configuration management policies for large buyers. In 2010 there are more than 60 metal mount specific RFID tags. The financial service industry, including Wells Fargo, bank of America, Morgan Stanley, Citigrop, Fidelity and others are purported to have tagged more than one million assets. At the same time new integrated circuits (ICs) were introduced by Alien, Impinj and NXP (formerly Philips) which proved much better performance and the IT Asset Tracking application exploded. The largest adopter to date appears to be Bank of America and Wells Fargo - each with more than 100,000 assets across more than a dozen data centers. In Feb 2008, Emirates Airline started a trial of RFID baggage tracing at London and Dubai airports. BGN has launched two fully automated Smartstores that combine item-level RFID tagging and SOA to deliver an integrated supply chain, from warehouse to consumer.

Logistics and transportation are major areas of implementation for RFID technology. For example, yard management, shipping and freight and distribution centers are some areas where RFID tracking technology is used. Transportation companies around the world value RFID technology due to its impact on the business value and efficiency. The North American railroad industry operates an automatic equipment identification system based on RFID. Locomotives and rolling stock are equipped with two passive RFID tags (one mounted on each side of the equipment); the data encoded on each tag identifies the equipment owner, car number, type of equipment, number of axles, etc. The equipment owner and car number can be used to derive further data about the physical characteristics of the equipment from the Association of American Railroads' car inventory database and the railroad's own database indicating the lading, origin, destination, etc. of the commodities being carried.

In January 2005, Wal-Mart required its top 100 suppliers to apply RFID labels to all shipments. To meet this requirement, vendors use RFID printer/encoders to label cases and pallets that require EPC tags for Wal-Mart.

These smart labels are produced by embedding RFID inlays inside the label material, and then printing bar code and other visible information on the surface of the label. In October 2005 the University of Arkansas' Information Technology Research Institute released a report on its preliminary study of the impact of RFID on reducing retail out-of-stocks and concluded that RFID reduced OOS by 21% over non-RFID based stores. In October 2007 Wal-Mart announced new focus areas for its RFID implementation.

1) Shipments going to Sam's Club
2) Promotional displays and products going to Wal-Mart stores
3) tests to see RFID's impact in improving category management in select areas.

In February 2009 Procter & Gamble stated it was ending its promotional program with Wal-Mart after Procter & Gamble "validated" benefits of the RFID program in merchandising and promotional displays. This implied Wal-Mart was not acting on the information to improve store execution.

WalMart, Target, BestBuy, and other retailers have discovered that RFID technology can keep inventories at the optimal level, reduce out-of-stock losses, limit shoplifting, and speed customers through check-out lines. Almost from the beginning, TI was there: helping establish standards; supporting the RFID supply chain of inlay and label manufacturers; and consistently applying leading-edge semiconductor technology to the core of RFID, the transponder. TI is unique among RFID semiconductor suppliers because it exercises complete design and manufacturing control over the entire transponder - a semiconductor chip and antenna. This value-add enables TI to offer a wider range of innovation whether it be in creative form factors, more attractive credit/debit cards, or technical issues such as read range and reliability. Innovation has been the touchstone of TI's long history in

RFID. It co-invented the smart tag, pioneered creative form factors such as key fobs, and executes its own antenna design.

The DoD requirements for RFID tags on packages is prescribed in the Defense Federal Acquisition Regulations Supplements (DFARS) 252.211-7006. Positioning of the tag needs to be completed in accordance with the clause and definitions in MIL STD 129 and as of 1 March 2007, EPC Global tags must comply with EPCglobal Class 1 Generation 2 specification. Another usage is visible. School authorities in the Japanese city of Osaka are now chipping children's clothing, back packs, and student IDs in a primary school. A school in Doncaster, England is piloting a monitoring system designed to keep tabs on pupils by tracking radio chips in their uniforms. St Charles Sixth Form College in west London, England, started September, 2008, is using an RFID card system to check in and out of the main gate, to both track attendance and prevent unauthorized entrance. As is Whitcliffe Mount School in Cleckheaton, England which uses RFID to track pupils and staff in and out of the building via a specially designed cards.In the Philippines, some schools already use RFID in IDs for borrowing books and also gates in those particular schools have RFID ID scanners for buying items at a school shop and canteen, library and also to sign in and sign out for student and teacher's attendance.

RFID technologies are now also implemented in end-user applications in museums. An example was the custom-designed temporary research application, "eXspot," at the Exploratorium, a science museum in San Francisco, California. A visitor entering the museum received an RF Tag that could be carried as a card. The eXspot system enabled the visitor to receive information about specific exhibits. The collected information could be retrieved at home from a "personalized" website keyed to the RFID tag. Over the past two decades, barcodes have made themselves useful everywhere from factory floors to neighborhood supermarkets. In manufacturing, few applications of IT are as popular as barcodes. They are universally appreciated for having improved data input productivity as well as data quality over manual keyboarding, and, in manufacturing, this past success may be the greatest barrier to RFID. The best opportunities for RFID are where barcoding is so pervasive that production operators spend a perceptible fraction of their time scanning barcodes. Such a situation is found in computer assembly, for pick validation and component serialization, and a switch to RFID in this context can be the vehicle to help the technology cross the chasm from a sprinkling of pilot systems to mass adoption.

RFID tags are often a replacement for UPC or EAN barcodes, having a number of important advantages over the older barcode technology. They may not ever completely replace barcodes, due in part to their higher cost and the advantage of multiple data sources on the same object. The new EPC, along with several other schemes, is widely available at reasonable cost. The storage of data associated with tracking items will require many terabytes. Filtering and categorizing RFID data is needed to create useful information. It is likely that goods will be tracked by the pallet using RFID tags, and at package level with Universal Product Code (UPC) or EAN from unique barcodes.

Standards are a complex game played by different rules in different countries and industries, with winners that are difficult to predict. The importance of standards, however, varies with the application. The Hong Kong Airport authority, for example, would like to use RFID luggage tags, but it would be useless unless other airports did too, with compatible tags and readers. A consensus on standards is essential for this application. For applications that are local to a manufacturing shop floor, on the other hand, hardware that follows standards may be cheaper and easier to get, but nonstandard hardware. In manufacturing, the absence of a consensus on standards may be an excuse to postpone the implementation of RFID, but it is not the kind of technical show stopper that it is for airports. There is presently no change to the existing requirement to have both a bar code label and human readable information provided on each shipment. This redundancy protects Wal-Mart from risks from the new technology, and ensures that all existing processes at the company do not need to change, but it does put an additional burden on the suppliers that are looking to deploy RFID.

The US Department of Defense (DoD) buys $24 billion dollars worth of packaged products annually, from more than 40,000 suppliers. The US DoD has been using RFID in logistics since Gulf War I. The huge volumes of materials that need to be transferred in a short period of time, with a high degree of precision in location, are a challenge for any technology. While the Wal-Mart mandate is unfunded, the DoD's comes with supplier support. For manufacturers, it remains an external mandate, not directly related to internal, and its main short term result may be more slap-and-ship.

The mandates from Wal-Mart and the DoD so far have not had this effect, as is evidenced by manufacturers resorting to slap-and-ship. While the result may be of value to Wal-Mart and the DoD, it is only a cost of doing business for the manufacturers. The current growth in the RFID industry is fueled by mandates from retailers and supermarkets that see cost-savings largely through labor reduction at the Shipping and Receiving docks. The cost justification, while obvious to retailers, is not for their manufacturing suppliers, and even large contract manufacturers perceive RFID technology as only cutting into margins.

The Slap-and-Ship approach to meeting customer mandates often involves the outsourcing of RFID compliance. In this model, the manufacturer pays a third party to ensure that shipments meet the RFID customer mandates. The third party may be a distributor or logistics provider that is already in the supply chain, or it may be a company that is contracted just for the compliance activities. The processing of the manufacturers products happens as follows:

1. The manufacturer ships finished goods, to the slap-and-ship service provider.
2. The service provider does the following:
   a. Break the shipment down to have access to each individual carton.
   b. Tag cartons per the customer's specifications.
   c. Rebuild pallets.
   d. Tag pallets per the customer's specifications.
   e. Ship tagged cartons and pallets to the customer
   f. Issue Advanced Shipment Notification (ASN) or other transactional information as required

Slap-and-ship requires minimal internal training and investment for hardware or software. It is a fast, pay-as-you-go way to satisfy inconsistent and evolving mandates from multiple customers. Beyond the short term, however, this approach has the following disadvantages:

1. It affords no opportunity for internal use of the information.
2. It involves no learning of the technology in the supplier organization.
3. It involves an additional link in the supply chain, which translates to more transportation and handling.

Most RFID applications in *manufacturing* are centered on the fully automatic identification of objects that may not be in the line of sight of the readers. They improve shopfloor inventory tracking and automate warehouse operations, including shipping / receiving.

Over the past few years, documented examples include the following:

* *Toyota (South Africa)* : Carrier tagged to streamline manufacturing and vehicle tracking. The tags are intended to remain with the vehicle throughout its life and hold its maintenance history.
* *Harley Davidson* : Process automation by tagging bins carrying parts to provide instructions to employees at each stage of the process.
* *Johnson Controls* : Tracking of car and truck seat through the assembly process.
* *TrenStar* : Beer keg tracking to improve demand forecasts and improving efficiency.
* *International Paper* : Paper roll tracking at for reduction of lost or misdirected rolls.
* *The Gap* : Denim apparel tracking to improve customer service through better inventory management.
* *Raxel* : Tagging reusable plastic biohazard containers to avoid contamination.
* *Michelin* : Tire tagging to comply with the TREAD act and recall management.

In 2002, Yoshikazu Suzuki, at Omron's Ayabe plant near Kyoto, developed what he called a "digital yatai" for nonrepetitive production of application-specific systems for displacement measurement, photoelectric sensing, film thickness measurement, digital fine scopes, or board solder inspection. Given that every assembly job in the digital yatai is custom, the operator cannot memorize sequences of tasks and relies on the visual instructions at every step. The transition between screens of instructions is triggered by sensors detecting the completion of each step, and the selection of parts and tools for each step is validated through auto-ID technology, using barcodes on parts and RFID tags on tools and fixtures. The system uses

PowerPoint to display instructions, a PLC and a DeviceNet network to communicate with sensors, barcode scanners, RFID readers and validation switches, with wireless interfaces between the work station and the shelves. Unlike most reported RFID applications in manufacturing, this one is not in production control or logistics but in the heart of the production process itself.

Barcodes have been used for this purpose, but, in computer assembly, barcode reading still involves either an operator moving the part under a reader or taking a mobile reader to the part. Because it involves extra labor, pick validation through barcodes does not fully qualify as mistake-proofing: in a crunch, it will be disabled to increase output. True mistake-proofing is only achieved with devices that are deeply embedded within the production process and require no additional labor. As a result, there is no short-term benefit to be gained from bypassing these devices. Satisfying this condition makes RFID a qualitatively superior solution to this problem. At a supermarket just off the line, a "water spider" picks kits of parts for and sequences these kits on a gravity flow rack a few minutes before they are used. The challenge here is to make sure that the water spider accurately picks the parts for each kit, and delivers the kits to the flow rack in sequence.

Retailer's initiatives to use RFID in an "open" supply chain application are a big reason why the technology is more prevalent in the industry today. But it's not the only reason for increased adoption. Several important factors came together around the same time. One factor is the advance in Ultra High Frequency (UHF) RFID systems. UHF systems are able to deliver the read range needed for supply chain applications, such as scanning tags on products as pallets are moved through a dock door or scanning cases on a high shelf in a warehouse. Another factor was the efforts by the Auto-ID Center to develop a system that is low cost and based on open standards.

These are prerequisites for the use of RFID in open supply chains, where a company puts a tag on a product, and it's read by other companies in the supply chain.

The ubiquity and advance of the Internet is another important factor. The Auto-ID Center realized that the Internet could be used to enable companies to share information about the location of products within the supply chain. Before the Auto-ID Center proposed the EPCglobal Network, there was no automated way (other than EDI) for Company 1 to let Company 2 know that it has shipped something, and there was no way for Company 2 to let Company 1 know that the product has arrived. Many companies use phones, faxes, email and spreadsheets to communicate with each other regarding shipment status. There are many other non-supply chain applications for RFID technology. These applications continue to advance along with supply chain applications. RFID technology continues to advance, to improve and to find new areas of application.

The advent of Radio frequency identification (RFID) promises great increase in efficiency in supply-chain management among other things. These benefits come for a price though. Concerns over adverse effect of RFID on the envionment have sparked protest from different groups. RFID tags are small objects attached to or built into products or other hosts that require identification, through a system of remote readers and data processors, similar to the bar-code system in use. While the largest application is the tracking of manufactured goods, they are also used for pet identification, medical prescriptions for the visually-impaired, bookstore / library book tracking, building access control, credit cards, badges, and in the future, for patient identification, intelligent traffic management, and are even envisioned for human implantation.

The pace of RFID implementation has received such a boost from its use by a well-known retailer that it is soon becoming the mantra for supply chain and inventory managers. Hectic activity is underway to realize the potential of RFID item tracking: that of efficient inventory control and, ultimately, cost benefits. While evolution of technical standards naturally ensued, ethical / environmental standards are taking much slower to evolve. It is true that companies have of late started to make products that are less harmful to the environment, or use less energy in their manufacture, or implement recycling policies sincerely. As with any other such issue, environmentalists have to cry hoarse before the powers that be hear them. There are many concerned groups asking and/or suggesting RFID tag manufacturers and users concerned to account for the direct and indirect impacts on the environment. Following are some cases of proactive agencies looking over RFID deployment.

Back in June 2000, the European Commission adopted a proposal for a Directive on Waste Electrical and Electronic Equipment (WEEE) and a proposal for a Directive on the restriction of the use of certain hazardous substances in electrical and electronic equipment. In 2005, the British Government has made an announcement on the timetable and policy leading to implementation of the WEEE Directive in the UK. In 2004, the US Environmental Protection Agency (EPA) and the Office of the Federal Environmental Executive (OFEE) co hosted a workshop on the impact of tags on the reuse and recycling of packaging materials. California based EoCycle electronic waste recyclers have a zero export policy, and has the capacity to break up and re-cycle e-waste into 13 material streams, and feed it back to local California businesses to make new products. Their mission is "to eradicate the improper disposal of obsolete electronic equipment from the corporate and institutional sectors in the most ethical, cost effective, and accountable manner.

However, it's possible to question whether all the uses are necessary at all for the purpose they are now serving in supply chain and inventory management of product retailers, arguably the largest sector of RFID use? Surely it must be possible to manage inventory and ensure proper checkout of goods from stores through other advancing technology. Wasn't barcode system serving the purpose? Severe anti-dumping laws are enforced in some countries, with little or no considerations for global consciousness and responsibility. The stage is here when the complete onus is on the manufacturer to exercise restraint and responsibility, guided by serious legislation. What we need is more laws and complete governmental will to implement them humanely, as well as continued awareness and action campaigns in all countries.

Perhaps the electronics manufacturers in general and RFID tag users in particular can take a leaf out of European car makers' books: by 2015, European law (End-of-Life Vehicles Directive) that becomes effective come January, will mandate that 95% of each automobile should be recycled. What's heartening in this nature-friendly mandate is the assurance that the reduction in potential waste to an astonishing 5% is achievable sans technology breakthroughs, costly or otherwise. Sanity and wisdom indicate that biodegradability should, and will replace all else as the corporate mantra in the near future.

Appendices

APPENDIX 1 : RFID GLOSSARY

1. Amplitude Modulation (AM)

A means of communication between an RFID tag and a reader; the data is contained in changes in the amplitude of the carrier wave sent out by the reader.

2. Antenna (RFID Tag Antenna)

The antenna in an RFID tag is a conductive element that permits the tag to exchange data with the reader. Passive RFID tags make use of a coiled antenna that can create a magnetic field using the energy provided by the reader's carrier signal.

3. Anti-collision

Anti-collision refers to different ways to keep radio waves from one device from interfering with radio waves from another device. RFID readers may make use of anti-collision algorithms to enable a single reader to read more than one tag in the reader's field.

4. Back scatter (RFID back scatter)

RFID tags sometimes make use of a method of communication called back scatter. Tags using back scatter technology reflect the reader's signal right back, modulating the signal to transmit data.

5. Bi-directional

RFID tags that are bidirectional are able to operate (that is, be read or be written to) from the front or from the back.

6. Chipless RFID tag

A chipless RFID tag (also known as RF fibers) is one that does not make use of any integrated circuit technology to store information. The tag uses fibers or materials that reflect a portion of the reader's signal back; the unique return signal can be used as an identifier.

The fibers are shaped in different ways; thin threads, fine wires or even labels or laminates. At volume, they range in cost from ten cents to twenty-five cents per unit.

7. Closed Systems (or Closed Loop Systems)

This term refers to RFID control systems in which the tracked objects never leave the company or organization. All of the data related to the object is stored in one place; the complete history of the object is readily available. The usual concerns about RFID standards do not apply, since the object tracked does not leave the system.

8. Code plate (RFID code plate)

Another term for an RFID tag. An RFID tag is a microchip combined with an antenna in a compact package; the packaging is structured to allow the RFID tag to be attached to an object to be tracked. RFID tags can be very small - the size of a large rice grain. Others may be the size of a small paperback book.

9. Contactless Credit Cards

A contactless credit card is different from an ordinary credit card. A contactless credit card uses RFID technology to store information about the account and to transfer it to the merchant. Standard credit cards carry data on a magnetic strip; when the card is swiped against a reader, the data is transferred.

10. Contactless smart card

The term Contactless smart card refers to identification cards (for example, some credit cards) that do not need to make contact with the reader to be read, or swiped in a special slot. This capability is implemented using a tiny RFID tag in the card; the intent is to provide the user with greater convenience by speeding checkout or authentication processes.

11. Continuous Wave (CM)

It is a mean of communication between an RFID tag and a reader; the data is contained in pulses of the carrier wave sent out by the reader.

12. Electromagnetic Interference (EMI)

A practical concern about the use of RFID arises in environments where radio waves of one device can distort or interfere with the radio waves of another device. Factories have many sources of radio waves, including wireless computers and electromechanical devices.

13. Electronic Article Surveillance (EAS)

This system uses tags that can be disabled by authorized agents. Sensors are placed at exit points; if the sensor detects a tagged object with an active tag, an alarm goes off.

14. Electronic Product Code (EPC)

The Electronic Product Code is a set of digits intended to complement barcodes like the Universal Product Code. An EPC is segmented, with digits that identify the manufacturer, the product category and the individual item.

15. Error Correcting Protocol

It is an Error correcting code.

16. Excite

The tag reader transmits radio frequency energy to

stimulate a passive RFID tag (to provide power to transmit its data back); the tag reader is said to excite the RFID tag.

17. Factory Programming

When the information on an RFID tag has been set during the manufacturing process, the resulting read-only tag is said to have factory programming.

18. Field Programming

Field programming refers to the process of placing data on an RFID tag after the manufacturing process has been completed. Field programming is usually performed before the tag is installed on the product or object.

Some field-programmable RFID tags have two forms of memory; a programmable chip and a form of read-only memory containing a unique serial number imposed at the factory.

19. Frequency Hopping

Frequency hopping is a technique used to keep two or more RFID readers from interfering with each other while reading RFID tags in the same area.

20. Frequency Modulation (FM)

A means of communication between an RFID tag and a reader; the data is contained in changes in the frequency of the carrier wave sent out by the reader.

21. Frequency Shift Keyed Modulation (FSKM)

A means of communication between an RFID tag and a reader; the data is contained in changes between the two frequencies of the carrier wave sent out by the reader.

22. Harvest

A passive RFID tag is said to harvest energy from an RFID reader antenna; passive tags have no energy source of their own, and they take what little they need from the reader.

23. Inductive Coupling

Inductive coupling is the transfer of energy from one circuit (such as a conductive antenna and associted circuitry) to another by means of mutual inductance between the two circuits.

24. Modulation

Modulation refers to the process by which the RFID tag changes the carrier signal of the reader antenna to convey information.

25. Passive RFID Tag (or Passive Tag)

A passive tag is an RFID tag that does not contain a battery; the power is supplied by the reader.

26. Phase Modulation (PM)

A means of communication between an RFID tag and a reader; the data is contained in changes in the phase of the carrier wave sent out by the reader.

27. Pulse Duration

A means of communication between an RFID tag and a reader; the data is contained in duration of pulses of the carrier wave sent out by the reader.

28. RF Fibers

RF fibers, also known as chipless RFID, is a kind of RFID tag that does not make use of any integrated circuit technology to store information. RF fibers can be used in more environments than RFID tags with electronic circuitry. They tend to work over a wider temperature range; these tags also are less sensitive to RF interference.

29. RFID Reader

An RFID reader is a device that is used to interrogate an RFID tag. The reader has an antenna that emits radio waves; the tag responds by sending back its data.

30. Smart label (RFID Smart Label)

A barcode label that contains an RFID transponder is

called an RFID smart label. Since the label is able to store information (like a serial number) and communicate with an RFID reader, it is considered "smart."

31. RFID tag

An RFID tag is a microchip combined with an antenna in a compact package; the packaging is structured to allow the RFID tag to be attached to an object to be tracked. "RFID" stands for Radio Frequency Identification.

32. Smart cards

Smart card is another way to refer to a contactless smart card. This term refers to identification cards (for example, some credit cards) that do not need to make contact with the reader to be read, or swiped in a special slot.

33. RFID Tag Collision

Tag collision in RFID systems happens when multiple tags are energized by the RFID tag reader simultaneously, and reflect their respective signals back to the reader at the same time. This problem is often seen whenever a large volume of tags must be read together in the same RF field. The reader is unable to differentiate these signals; tag collision confuses the reader.

34. Transponder for RFID

An RFID transponder is a special kind of radio transmitter and receiver. It is activated when it receives a signal of a specific kind. RFID transponders are present in smart cards and Radio Frequency Identification tags.

APPENDIX 2 : NFC TERMINOLOGY AND GLOSSARY

Terminology

1. **NDEF** - NFC Data Exchange Format - standard exchange formats for URI, Smart Posters, other
2. **RTD** - Record Type Definition - An NFC-specific record type and type name which may be carried in an NDEF record
3. **NDEF message** - Basic message construct defined by this specification. An NDEF message contains one or more NDEF records
4. **NDEF record** - Contains a payload described by a type, a length, and an optional identifier
5. **NDEF payload** - The application data carried within an NDEF record

Glossary:

1. **3GPP:** 3rd Generation Partnership Project: worldwide cooperation between companies in order to drive mobile communication.
2. **AFC:** Automated Fare Collection: A billing concept for public transport operators based on a check-in/ check-out system.
3. **APDU Application Protocol Data Units:** Data Exchange format specified in ISO7896 to exchange data between a reader and smartcard chips
4. **BOM:** Bill of Materials: the cost of the material of a device being manufactured (no IP or labour cost)
5. **BIP Bearer Independent Protocol:** Used in terms of SAT. Applications running on the SIM could make use of various data bearers like BT, SMS or data if supported by the SIM.
6. **c-less:** Contactless
7. **CAGR:** Common Average Growth rate: usually mention in regard to business cases

8. **CA:** Certificate Authority
9. **CASD CA's SD**
10. **CC:** Common Criteria: A common agreement on security standards in IT infrastructures.
11. **CCM** Card Content Management
12. **CP Logical Link Control Protocol:** Protocol used in NFC peer to peer transactions for a smooth establishment of communication. (e.g. handles Initiator + Target Configuration)
13. **EAL:** Evaluation Assurance Level
14. **EE:** Execution Environment
15. **EPC:** Electronic Product Code. A data format specification for unique identification of products
16. **EMV:** EMV = Europay, Mastercard, Visa. Agreement between credit card companies on a standardized protocol for transaction between a credit card and a card reader. (Mainly security focused).
17. **Felica:** A product of Sony: This smartcard product is not compatible to ISO14443 but implements a proprietary radio modulation.
18. **GSMA:** GSM Association: Association of about 700 mobile network operators (MNOs) in 218 countries around the world HIC *: Host Controller Interface*: A specified interface for the NFC Chip for the communication with the host controller/baseband controller of the device.
19. **GP:** Global Platform: An industry standard defining the APIs, interfaces and commands for security managing applications on a smartcard
20. **IC:** Integrated Circiut
21. **IOS:** *independent organizations of sales* that sell equipment and services to merchants
22. **ID Issuer Domain:** the security domain of the card issuer.

23. **IOT:** Internet of Things: A term used to describe ecosystems based on objects being tagged with RF transponders
24. **JCOP Applet:** A java application running on a java card OS based smartcard; also referred as "Applet" (without JCOP) in terms of NFC
25. **Java Card OS:** A java based operating system for sophisticated smartcards.
26. **JCOP:** Java Card Open Platform: Cards based on Java Card OS implementing the global platform specifications for management applications on the smartcard
27. **JSR257:** A Java API standard defined to make use of the capability of the NFC functionality of the handset.
28. **J2ME:** Java Platform Micro Edition: Java Implementation for mobile device
29. **JSR177** see also SATSA
30. **Mifare:** A product of NXP Semiconductors. It is a smartcard based on ISO14443A providing up to 4 KB of memory on the card.
31. **MNO:** Mobile Network Operator: The mobile network operator provides a platform for secure application for NFC if having the UICC as the SE
32. **MA Multi Application;** usually mention in terms with smart cards and Global Platform (GP).
33. **ME: Mobile Equity:** mobile device providing a SIM card slot
34. **NFC: Near Field Communication:** A communication technology standardized in ISO 18092 and ETSI allowing bidirectional communication between two devices based on RF technology. This is also refereed as the peer to peer (P2P) mode of NFC devices.
35. **NFCC: Near Field Communication Controller:**

Analog/Digital Frontend IC inthe device to enable it for NFC transactions.

36. **NFCIP-1:** Standard for NFC (peer to peer mode) standardized in ISO18092 and ECMA 320/340.
37. **NDEF: NFC Data Exchange Format:** A data format to store information like URL, SMS or plain text in a data container that could be stored on an NFC tag or exchanged over NFC (peer to peer mode). One NDEF can contain multiple Record Type Definition (RTD) and therefore is also referred as NDEF data container.
38. **NFC Device:** A device featuring different operating modes based on RF technology. There are three modes: tag emulation (PICC), NFC peer to peer (NFC) and reader/writer mode (PCD). NFC Devices must implement at least mandatory parts of the NFC Protocol Stack. The NFC Devices' operating modes are defined by the NFC Forum.
39. **OTA: Over the Air:** The possibility to send and receive data to/from a device in distributed environment. In GSM networks data connection or SMS could be used to do so.
40. **PTO:** Public Transport Operator
41. **POS:** Point of Sale
42. **PC/SC Personal Computer/Smart Card:** A defined Interface (Software) between a personal computer and a smart card. Used for communication with a smartcard reader.
43. **PCD: Proximity coupling device:** a transmitter that can read tags based on ISO14443 (PICC). The reader emits an electromagnetic field that powers a tag/ transmitter by inductivity. Communicates with PICC using load modulation scheme.
44. **PICC: Proximity inductive coupling card:** a transponder that can be read or written by a

proximity reader. Theses tags are based on the ISO14443 standard. Such tags do not have a power supply like a battery, but are powered by the electromagnetic field of the reader (PCD)

45. **PKI:** Public Key Infrastructure
46. **RFID:** Radio Frequency Identification: technology used to identify objects carrying RF transponders. NFC and RFID technology have overlapping standards and terminology.
47. **RTD:** Record Type Definition: Specification on how information like URI, SMS or plain text are encoded to fit into NDEF data containers.
48. **SAM:** Secure Access Module: module for the purpose of securely storing data, also referred as SE
49. **SD:** Security Domain: Part of a Global Platform (GP) compliant smartcard to hold the application of one Service Provider (SP)
50. **SE:** Secure Element: a Trusted Environment for storing sensitive data or applications.
51. **SCWS Servlet** An application running in the servlet container of a Smart Card Web Server; also referred as "Servlet" (without SCWS) in terms of NFC.
52. **SIM:** Subscriber Identity Module: A smart card in a special form factor used in a mobile phone to authenticate the user against the network.
53. **SP:** Service Provider: an entity in the NFC ecosystem providing application. These application do not need to be based on a SE.
54. **SCWS Smart Card Web Server:** A web server running on a smartcard
55. **SAT SIM Application Toolkit:** Defined Interface/API between SIM and Mobile Equity (ME) for applications running on the SIM Card
56. **SATSA:** Security and trust service API: A Java API

standard defined in JSR177 allowing a J2ME Application to access the SE of a mobile device

57. **SVA:** Store Value Account: an account containing sensitive data like money or points to be used in a smartcard based infrastructures.
58. **SWP: Single wire protocol:** a connection between the NFC Chips and the UICC to be used as the secure element; also referred as the USIM
59. **T=CL** Transport Protocol for contactless communication in the field of smartcards (related to bit/encoding level on RF layer)
60. **T=1 or T=0Transport** Protocol for contacted communication in the field of smartcards (related to bit/encoding level on physical layer)
61. **TPM** : Trusted Platform Manger; see TSM
62. **TSM:** Trusted Service Manager: A trusted entity in an NFC ecosystem managing the SE and acting as a moderator and resolving the m-to-n between MNOs and SPs.
63. **TTP:** Trusted Third Party: A trusted entity in a distributed environment. The TTP is the Certificate Authority of a Public Key Infrastructure. The TTP is able to issue self-signed certificates.
64. **UICC:** UMTS integrated circuit card: SIM card for UTMS networks
65. **USIM:** UMTS SIM; see UICC
66. **VAS** Value added service

Bibliography

References and Notes

1. Hacking Exposed Linux: Linux Security Secrets & Solutions (third ed.). McGraw-Hill Osborne Media. 2008. p. 298. ISBN 978-0-07-226257-5. http://books.google.com/books?id=f5Vz08spzw8C&printsec=frontcover.
2. Cardullo et al. "Transponder apparatus and system" U.S. Patent 3,713,148 (1973)
3. Genesis of the Versatile RFID Tag, RFID Journal http://www.rfidjournal.com/article/view/392/1/2
4. Landt, Jerry (2001). "Shrouds of Time: The history of RFID" (PDF). AIM, Inc.. http://www.transcore.com/pdf/AIM%20shrouds_of_time.pdf.
5. Charles A. Walton "Portable radio frequency emitting identifier" U.S. Patent 4,384,288 issue date May 17, 1983
6. http://www.rfid-handbook.de/rfid/index.html
7. http://www.st.com/internet/mcu/subclass/1200.jsp
8. http://www.rfidtribe.com/index.php?option=com_content&task= view&id=115&Itemid=102
9. "Savi is a Leader in Wireless Tracking Solutions for Managing Assets, Inventory and Shipments". Savi. http://www.savi.com/.

10. "ODIN Technologies-ODIN rfid software and solutions". Odintechnologies.com. http://www.odintechnologies.com/.
11. Airbus Selects ODIN technologies as Exclusive Worldwide RFID Partner http://www.industrial-embedded.com/news/db/?6892
12. Ants' home search habit uncovered http://news.bbc.co.uk/1/hi/england/bristol/8011998.stm
13. RFID Tag Characteristics. http://www.zebra.com/id/zebra/na/en/index/rfid/faqs/rfid_tag_characteristics.html
14. "Hitachi's RFID powder freaks us the heck out". Engadget. http://www.engadget.com/2007/02/14/hitachis-rfid-powder-freaks-us-the-heck-out.
15. TFOT (2007). "Hitachi Develops World's Smallest RFID Chip". http://thefutureofthings.com/news/1032/hitachi-develops-worlds-smallest-rfid-chip.html.
16. Imagineering (2007). "Optically Powered RFID Tags and Optical Tag Readers". http://www.imagineering-ezine.com/e-zine/OPID.html.
17. RFID World http://www.rfid-world.com/products/207200945
18. Tay, Lay (2007-11-01). "HDB Introduces RFID Season Parking Ticket". RFID Asia. http://www.rfid-asia.info/2007/11/hdb-introduce rfid-season-parking.htm.
19. "UPM Raflatac Supplies RFID Inlays for Moscow Metro Contactless Tickets". RFID Journal. http://www.rfidjournal.com/article/articleview/3092/1/79/.
20. m-Card http://mcard.metrobus.com/html/index.htmlMetrobus
21. https://www.prestocard.ca/StaticContent/Gtafs
22. "PRESTO - Convenience And Simplicity". News.ontario.ca. 2010-05-10. http://www.news.ontario.

ca/mto/en/2010/05/presto---convenience-and-simplicity.html.

23. "Tampereen Kaupunkiliikenne. Liikelaitoksen historiaa". http://www.tampere.fi/tkl/virstanpylvaat.html.
24. "Monedero card". http://www.monedero.com.ar/.
25. "Architecture/Infrastructure by Bank Systems & Technology". Banktech.com. 2010-04-19. http://www.banktech.com/architecture-infrastructure.
26. Webster (2008-02-13). "BAA and Emirates test new baggage tagging technology". Jetvine. http://jetvine.blogspot.com/2008/02/baa-and-emirates-test-new-baggage.html.
27. Beth Bacheldor (2007-10-18). "Agricultural Company Tracks Equipment Loaned to Farmers". RFID Journal. http://www.rfidjournal.com/article/articleview/3690/1/1/.
28. Mary Catherine O'Connor (2007-02-13). "Purdue Moving OxyContin RFID Pilot to Full Production". RFID Journal. http://www.rfidjournal.com/article/articleview/3043/1/1/.
29. Rhea Wessel. "Berliner Wasserbetriebe Gets RFID Tagging Project Flowing". RFID Journal. http://www.rfidjournal.com/article/view/3110/2
30. "AEI technology". Softrail. http://www.aeitag.com/aeirfidtec.html.
31. USDA Bets the Farm on Animal ID Program http://www.thenation.com/doc/20071231/pentland_gumpert
32. RFID's reduction of Out-of-Stock study at Wal-Mart, RFID Radio http://www.rfidradio.com/?p=11
33. RFID's Second Wave, BusinessWeek http://www.businessweek.com/technology/content/aug2005/tc2005089_4131_tc_215.htm

34. ClearCount Medical Solutions - SmartSponge System®. http://www.clearcount.com/smart-sponge-system.html
35. Bacheldor, Beth (2008-01-11). "Sam's Club Tells Suppliers to Tag or Pay". http://www.rfidjournal.com/article/articleview/3845/1/1/.
36. ""Looking Back at the Wal-Mart RFID Time Line". 2009-03-24. http://www.scdigest.com/assets/On_Target/09-02-23-1.php?cid=2275.
37. DFARS 252.211-7006 clause (a) (2).
38. "Keeping Track of Promotion Progress: How Marketing Will Become the Greatest Advocate of RFID". Consumer Goods Technology. http://www.consumergoods.com/ME2/dirmod.asp?sid=&nm=&type=Publi shing&mod= Publications%3A%3AArticle&mid=8F3A7027421841978F18B E895 F87F791 &tier= 4&id=07CA1C544D3E4 FD1916 C7A6D263891 3E.
39. Radio Frequency Identification: An Introduction for Library Professionals. Alan Butters. Australasian Public Libraries v19.n4(2006) pp.2164-174.
40. "The State of RFID Applications in Libraries." Jay Singh et al. Information Technology & Libraries no.1(Mar.2006) pp.24-32.
41. Making RFID work: The World's largest University Library RFID Implementation, Center of Asia Business Cases, The University of Hong Kong, (09/451), 2009.
42. "Radio Frequency Identification." Rachel Wadham. "Library Mosaics" v14 no.5 (S/O 2003) pg.22.
43. AudioIndex - the Talking Library, http://www.umea.se/omkommunen/languages/inenglish/culturalactivitieslibraries/libraryservices/projects/library2007/audioindexthetalkinglibrary.4.64e618c10982eb6c427fff10897.html

44. Rahman, Rohisyam (2007-07-23). "Case Study: Malaysian Smart Shelf". RFID Asia. http://www.rfid-asia.info/2007/07/case-study-malaysian-smart-shelf.htm.

45. "RFID Poses No Problem for Patron Privacy." "American Libraries" v34 no11 (D 2003) pg.86.

46. Datatilsynet misfornøyd med nye pass – digi.no also in Biometric passport#Norway

47. "Contactless inlays from SMARTRAC ordered for US ePassport project". http://www.contactlessnews.com/2006/11/30/contactless-inlays-from-smartrac-ordered-for-us-epassport-project.

48. "United States sets date for E-passports". http://rfidjournal.com/article/articleview/1951/1/132/.

49. "Austrian Foreign Ministry -> Embassy -> Ottawa -> Passport". Bmeia.gv.at. http://www.bmeia.gv.at/en/embassy/ottawa/practical-advice/austrians-in-canada/passport.html.

50. Schoolchildren to be RFID-chipped. http://networks.silicon.com/lans/0,39024663,39122042,00.htm

51. Schoolkid chipping trial 'a success'. http://www.theregister.co.uk/2007/10/22/kid_chipping_doncaster_go/

52. S. Hsi en H. Fait, "RFID enhances visitors Museum Experience at the Exploratorium," Communications of the ACM 48, 9 (2005): 60

53. Social Shopping in a Fully Enabled RFID Store, RFID Radio http://www.rfidradio.com/?p=20

54. Barcelona clubbers get chipped. BBC News, 29 Sep 2004

55. "VeriChip RFID Implants in Mexican Attorney General's Office Overstated". Spychips.com. 2004-11-29. http://www.spychips.com/press-releases/mexican-implant-correction.html.

56. High-tech cloning http://blogs.reuters.com/blog/2006/07/22/high-tech-cloning
57. Vericip hacked press release, Spychips http://www.spychips.com/press-releases/verichip-hacked.html
58. "Demo: Cloning a Verichip" http://cq.cx/verichip.pl.
59. "VeriChips Implanted at CityWatcher.com". Compliance and Privacy. http://www.complianceandprivacy.com/News-VeriChip-concern.asp.
60. CASPIAN Special Report, October 19, 2004: FDA Letter Raises Questions about VeriChip Safety, Data Security http://www.spychips.com/devices/verichip-fda-report.html
61. Rosenberg, Isaac B. (2008), Involuntary Endogenous RFID Compliance Monitoring as a Condition of Federal Supervised Release - Chips Ahoy?, 21, Yale Journal of Law & Technology, pp. 331, http://ssrn.com/abstract=1079106
62. Martein Meints (2007), D3.7 A Structured Collection on Information and Literature on Technological and Usability Aspects of Radio Frequency Identification (RFID), FIDIS deliverable 3(7), June 2007.
63. Paolo Magrassi (2001). "A World Of Smart Objects: The Role Of Auto Identification Technologies". http://www.gartner.com/.
64. Pete Harrison (2009). "EU considers overhauling rules for lost air luggage". http://www.reuters.com/article/rbssIndustryMaterialsUtilitiesNews/idUSLS63631320090728.
65. Fisher, Jill A. 2006. Indoor Positioning and Digital Management: Emerging Surveillance Regimes in Hospitals. In T. Monahan (Ed), Surveillance and Security: Technological Politics and Power in Everyday Life (pp. 77-88). New York: Routledge.
66. Fisher, JA; Monahan, T (2008). "Tracking the social

dimensions of RFID systems in hospitals.". International journal of medical informatics 77 (3): 176-83. doi:10.1016/j.ijmedinf.2007.04.010. PMID 17544841. http://torinmonahan.com/papers/Fisher_Monahan_RFID_IJMI.pdf.

67. "Electronic tags for eggs, sperm and embryos - life - 30 March 2005". New Scientist. http://www.newscientist.com/article.ns?id=dn7209.
68. Bacheldor, Beth. "RFID-enabled Handheld Helps Nurses Verify Meds." RFID Journal. July 2007.
69. "EdgeMagic To Help Medical Supply Company Comply | CYBRA Corporation". Cybra.com. 2009-03-23. http://www.cybra.com/press/202-em-article2.html.
70. 'Connor, Mary Catherine (March 18, 2009). Dash7 Alliance Seeks to Promote RFID Hardware Based on ISO 18000-7 Standard. RFID Journal LLC. http://www.rfidjournal.com/article/articleview/4696/1/1.
71. DASH7 Alliance home page http://www.dash7.org/
72. Roberti, Mark (2006-05-06). "A 5-Cent Breakthrough". RFID Journal. http://www.rfidjournal.com/article/articleview/2295/1/128/.
73. "Radio Silence". The Economist. 7 June 2007. http://economist.com/printedition/displaystory.cfm?story_id=9249278.
74. "Financial Services Technology Consortium - Projects". FSTC. 2008-12-05. http://www.fstc.org/projects/index.php?id=29.
75. http://www.bentsystems.com/rfidhistory.shtml
76. http://www.mbcestore.com/rfid/history.htm
77. http://en.wikipedia.org/wiki/RFID
78. http://www.andrew.cmu.edu/user/cjs/tech.html
79. http://en.wikipedia.org/wiki/Identification_friend_or_foe

80. http://www.aimglobal.org/technologies/rfid/resources/shrouds_of_time.pdf

81. http://www.rfidjournal.com/glossary/125

82. http://www.slais.ubc.ca/courses/libr500/04-05-wt2/www/T_Gnissios/history.htm

83. http://www.rfidtribe.com/index.php?option=com_content&view=section&id=6&Itemid=102

84. http://www.discoverrfid.org/how-it-works/history/development-of-rfid.html

85. http://www.discoverrfid.org/how-it-works/history/from-the-barcode-to-epc.html

86. http://www.discoverrfid.org/how-it-works/history.html

87. History of RFID by Dr. Jerry Landt (*TransCore's Chief Scientist and an Original Member of the Los Alamos Scientific Team that Developed RFID for the Federal Government and Cofounder of the Amtech technology).* http://www.aimglobal.org/technologies/rfid/resources/shrouds_of_time.pdf

88. RFID - The Early Years 1980-90 : http://members.surfbest.net/eaglesnest/rfidhist.htm

89. History from RFID Journal : http://www.rfidjournal.com/article/articleview/1338/1/129/

90. RFID Timeline from Texas Instrument : http://www.ti.com/rfid/timeline/timeline.shtml

91. White Paper & Timeline from Interaction Design Institute : http://people.interaction-ivrea.it/c.noessel/RFID/research.html

92. RFID / Wikipedia : http://en.wikipedia.org/wiki/RFID - version française http://fr.wikipedia.org/wiki/RFID

93. http://www.radio-electronics.com/info/wireless/radio-frequency-identification-rfid/development-history.php

94. "RCD Introduces its Worldwide Compatible RFID Tag

- Sun Microsystems is First to Implement". RCD Technology. http://www.rcdtechnology.com/rcd-introduces-its-worldwide-compatible-rfid-tag-sun-microsystems-is-first-to-implement/546/.

95. "What's New". Radio Frequency Identification (RFID). 4 April 2007. http://www.acq.osd.mil/log/rfid/index.htm.

96. Tedjasaputra, Adi (2006-12-11). "Putting RFID Network Security in Perspective". RFID Asia. http://www.rfid-asia.info/2007/02/putting-rfid-network-security-in.htm.

97. "RFID Privacy and Security". RSA Laboratories. http://www.rsasecurity.com/rsalabs/node.asp?id=2115.

98. "RFID chips can carry viruses". Ars Technica. http://arstechnica.com/news.ars/post/20060315-6386.html

99. "RFID Passports cracked. Easily, cheaply, and quickly". wired. http://blog.wired.com/sterling/2006/11/arphid_watch_fi.html.

100. "RFID Passports cracked through the mail". the register. http://www.theregister.com/2007/03/06/daily_mail_passport_clone/.

101. "FIPS-201, Personal Identity Verification (PIV) of Federal Employees and Contractors" (PDF). NIST. http://csrc.nist.gov/publications/fips/fips201-1/FIPS-201-1-chng1.pdf.

102. "Can Aluminum Shield RFID Chips?". RFID Shield. http://www.rfid-shield.com/info_doesitwork.php.

103. "Aluminum Foil Does Not Stop RFID". Omniscience is Bliss. http://www.omniscienceisbliss.org/rfid.html.

104. "System and Method for Providing a Low and Narrow-Profile Radio Frequency Identification (RFID) Tag". USPTO. http://appft.uspto.gov/netacgi/nph-Parser?Sect1= PTO2&Sect2=HITOFF&u= %2Fnetahtml%2FPTO%2Fsearch-adv.html&r=1&p= 1&f=G&l=

50&d=PG01&S1= 20100127876& OS= 20100127876&RS=20100127876.

105. Gilbert, Alorie (18 August 2003). "Privacy advocates call for RFID regulation". News.com. http://news.cnet.com/Privacy-advocates-call-for-RFID-regulation/2100-1029_3-5065388.html.

106. Markus Hansen, Sebastian Meissner: Identification and Tracking of Individuals and Social Networks using the Electronic Product Code on RFID Tags, IFIP Summer School, Karlstad, 2007, Slides. https://tepin.aiki.de/blog/uploads/20070807-ifip-hansen-meissner-tracking-rfid-epc.pdf

107. "Today at PC World". http://blogs.pcworld.com/staffblog/archives/000798.html.

108. http://blog.makezine.com/archive/2005/07/_defcon_rfid_wo.html. http://blog.makezine.com/archive/2005/07/_defcon_rfid_wo.html.

109. Katherine Albrecht, Liz McIntyre. "The METRO "Future Store" Special Report". Spychips. http://www.spychips.com/metro/overview.html.

110. Richard M Stallman. "The WSIS in Tunis". http://www.fsf.org/blogs/rms/entry-20060125.html.

111. K. C. Jones. "Free Software Advocate Attracts U.N. Security After Blocking RFID Tags". Infowars. http://www.infowars.com/articles/bb/rfid_stallman_blocks_rfid_tag_at_un_summit.htm.

112. FTC "Radio Frequency Identification: Applications and Implications for Consumers" (March 2005) http://www.ftc.gov/os/2005/03/050308rfidrpt.pdf

113. monochrom. "R F I D". http://www.monochrom.at/rfid/.

114. Greene, Thomas C. (2004). "Feds approve human RFID implants". http://www.theregister.co.uk/2004/10/14/human_rfid_implants/.

115. "Barcelona clubbers get chipped". news.bbc.co.uk

(BBC News). 2004-09-29. http://news.bbc.co.uk/go/pr/fr/-/2/hi/technology/3697940.stm.

116. Monahan, Torin and Tyler Wall. 2007. .Somatic Surveillance: Corporeal Control through Information Networks. Surveillance & Society 4 (3): 154-173

117. Reuters http://blogs.reuters.com/2006/07/22/high-tech-cloning/

118. Susan. "Using Radio Frequency Identification (Rfid) Technology In Humans In The United States For Total Control By David B. Smith A Grad". 74.125.155.132. http://74.125.155.132/scholar?q=cache:aO1IdzD4K0cJ:scholar.google.com/+%22david+b.+smith%22+rfid&hl=en.

119. Iain Thomson in San Francisco. "Hacker clones passports in drive-by RFID heist - V3.co.uk - formerly vnunet.com". V3.co.uk. http://www.v3.co.uk/vnunet/news/2235666/hackers-clones-passports-drive.

120. "Human chips more than skin-deep | Tech News on ZDNet". News.zdnet.com. http://news.zdnet.com/2100-9584_22-138001.html.

121. Katherine Albrecht, Liz McIntyre (2005). Spychips: how major corporations and government plan to track your every move with RFID. Thomas Nelson Inc. ISBN 1595550208. http://books.google.com/books?hl=en&lr=&id=YDxDuMYVJdcC&oi=fnd&pg=PR9&dq=RFID+Governm ent&ots=bnMUITNKNy&sig=EN-XmWBRttnJ5FbZx KOVS4PCieQ#v=onepage&q=&f=false.

122. RFID faq http://www.rsa.com/rsalabs/node.asp?id=2120#13

123. RFID Tag Characteristics. http://www.zebra.com/id/zebra/na/en/index/rfid/faqs/rfid_tag_characteristics.html

124. http://www.rcdtechnology.com/rfid-applications/success-stories/

125. http://nje-rfid.com/RFIDTechnical.htm
126. http://www.rfid-handbook.de/rfid/measurements.html
127. http://www.slais.ubc.ca/courses/libr500/04-05-wt2/www/T_Gnissios/implementation.htm#top
128. http://www.slais.ubc.ca/courses/libr500/04-05-wt2/www/T_Gnissios/libraries.htm#top
129. http://www.barcoding.com/barcode/barcode-faqs.shtml
130. http://www.rfidjournal.com/faq
131. http://www.libraryjournal.com/
132. http://www.slais.ubc.ca/courses/libr500/04-05-t2/www/T_Gnissios/biblio.htm#4
133. http://www.vpl.ca/
134. http://www.slais.ubc.ca/courses/libr500/04-05-wt2/www/T_Gnissios/biblio.htm#5
135. http://www.slais.ubc.ca/courses/libr500/04-05-wt2/www/T_Gnissios/costs.htm
136. How do I calculate ROI for RFID? SearchSAP.com http://searchsap.techtarget.com/resources/SAP-ROI-and-TCO
137. http://www.slais.ubc.ca/courses/libr500/04-05-wt2/www/T_Gnissios/benefits.htm
138. http://www.slais.ubc.ca/courses/libr500/04-05-wt2/www/T_Gnissios/problems.htm
139. Peter de Jager. (2005). http://www.technobility.com/
140. Guidelines. http://www.epcglobalinc.org/standards/implementation_guidelines/
141. http://www.infosciencetoday.org/tag/rfid-for-library
142. http://www.infosciencetoday.org/tag/rfid-library
143. Syed, Mohammad_Shahid. Use of RFID Technology in Libraries: An Automated Metheod of Circulation, Security, Tracking. http://ezinearticles.com/

144. Ayre, Lori Bowen, The Galecia Group (August 2004) Position paper: RFID and libraries. Retrived from http://www.galecia.com/weblog/mt/archives/cat_rfidandwireless.php

145. Berkeley Public Library (n.d.) Berkeley Public Library: Best Practices for RFID technology. Retrieved from http://berkeleypubliclibrary.org/BESTPRAC.pdf.

146. BIBLIOTHECA RFID Library Systems AG (2003) RFID Technology Overview Retrieved from http://www.bibliotheca-rfid.com

147. Boss. R. W. (2003). RFID technology for libraries [Monograph]. Library Technology Reports. November-December 2003.

148. Boss. R. W. PLA Tech Notes (May 14, 2004) RFID Technology for libraries. Retrieved from http://www.ala.org/ala/pla/plapubs/technotes/rfidtechnology.htm

149. FAQ RFID Journal (OnlineVersion) Retrieved from http://www.rfidjournal.com/article/articleview/207

150. Koppel, T. (March 2004). Standards in Libraries: What's Ahead: a guide for Library Professional about the Library Standards of Today and the Future. The Library Corporation. Retrieved from http://www.tlcdelivers.com/tlc/pdf/standardswp.pdf.

151. Molnar, D., Wagner, D. A. (June 2004). Privacy and security in library RFID: Issues, practices and architectures. Retrieved from http://www.cs.berkeley.edu~dmolnar/library.

152. Sarma, E. S. Weis, S. A., Engels, D.W. (November 2002). White paper: RFID systems, security & privacy implications. Cambridge, MA: Massachusetts Institute of Technology, AUTO-ID Center.

153. Allied Business Intelligence. (2002). RFID white paper. Oyster Bay, New York.

154. American Library Association. (2001). Code of Ethics

of American Library Association. Retrieved August 5, 2004 from http://www.ala.org/ala/oif/statementspols/codeofethics/codeethics.htm

155. Aarts, E., Harwig, R. Schuurmans, M. (2002). Ambient Intelligence. In P.J. Denning (Ed.), The invisible future, pp.235-250). New York: McGraw Hill.

156. Bednarz, A. (2004, May 3). RFID everywhere: From amusement parks to blood supplies. Network World Fusion. Retrieved August 4, 2004 from http://www.nwfusion.com/news/2004/0503widernetrfid.html.

157. Bednarz, A. (2003, November 3). Defense Department goes on offense with RFID. Network World Fusion. Retrieved August 4, 2004 from http://www.nwfusion.com/news/2003/1103forrester side.html

158. Berkeley Public Library. (n.d.). Berkeley Public Library: Best practices for RFID technology. Retreived August 5, 2004 from http://berkeleypubliclibrary.org/BESTPRAC.pdf.

159. Booth-Thomas, C. (2003, September 22). The See-it-all chip: Radio-frequency identification—with track-everything-anywhere capability, all the time—is about to change your life. Time, 162, pA8. Retrieved July 9, 2004 from Gale Group database.

160. Boss, R.W. (2003). RFID technology for libraries [Monograph]. Library Technology Reports, November-December 2003.

161. Bruening, P. (2004). Testimony presented at House Subcommittee on Commerce, Trade and Consumer Protection on July 24, 2004. Washington, DC. Retrieved August 4, 2004 from http://www.cdt.org/testimony/20040714bruening.pdf.

162. Cavoukian, A., Ph.D. (2004, February). Tag, you're it: Privacy implications of radio frequency identification

(RFID) technology. Toronto, Ontario: Information and Privacy Commissioner.

163. Cavoukian, A., Ph.D. (2004, June). Guidelines for using RFID tags in Ontario public libraries. Toronto, Ontario: Information and Privacy Commissioner.
164. Consumers Against Supermarket Privacy Invasion and Numbering (CASPIAN). (2003). Position statement on the use of RFID on consumer products. Available from http://www.privacyrights.org/ar/RFIDposition.htm
165. Dorman, D. (2002, September). New vendors heating up radio frequency ID market. American Libraries, 33(8). Retrieved July 9, 2004 from MasterFILE Premier database.
166. Dowbenko, U. (n.d.) VeriChip: RFID microchip implants for humans. Retrieved July 11, 2004 from http://www.conspiracyplanet.com/channel.cfm?channelid=74&contentid=900
167. EPC Global. (2003) Guidelines on EPC for Consumer Products. Retrieved July 11, 2004 from http://www.epcglobalinc.org/public_policy/public_policy_guidelines.html
168. Flagg, G. (2003, December). Should libraries play tag with RFID? American Libraries, 34(11), 69-71 Chicago: American Libraries.
169. Hesseldahl, A. (2004, July 29). A hacker's guide to RFID. Forbes.com. Retrieved August 6, 2004 from http://www.forbes.com/home/commerce/2004/07/29/cx_ah_0729rfid.html.
170. Garfinkel, S. (2001). Database Nation. Sebastopol, California: O'Reilly & Associates.
171. Garfinkel, S. (2002, October). An RFID bill of rights. MIT Technology Review. Retrieved July 8, 2004 from http://www.simson.net/clips/2002.TR.10.RFID_Bill_Of_Rights.htm

172. Givens, B. (2004, January 10). RFID implementation in libraries: Some recommendations for "Best Practices." Summary of presentation to the ALA Intellectual Freedom Committee of the American Library Association at ALA Mid-Winter, San Diego, California. Retrieved from http://www.privacyrights.org/ar/RFID-ALA.htm.

173. Koppel, T. (2004, March). Standards in Libraries: What's Ahead: A Guide for Library Professionals About the Library Standards of Today and the Future. The Library Corporation. Retrieved August 5, 2004 from http://www.tlcdelivers.com/tlc/pdf/standardswp.pdf.

174. Landt, J., Catlin, B. (2001). Shrouds of time: The history of RFID. Pittsburg, Pennsylvania: AIM, Inc. Retrieved August 3, 2004 from http://www.aimglobal.org/technologies/rfid/resources/shrouds_of_time.pdf.

175. McHugh, J. (2004, July) Attention, shoppers: You can now speed straight through checkout lines! Wired 12.07 (online version). Retrieved July 9, 2004 from http://www.wired.com/wired/archive/12.07/shoppers.html

176. Molnar, D., Wagner, D.A. (2004, June 8). Privacy and security in library RFID: Issues, practices and architectures. Retrieved July 30, 2004 from http://www.cs.berkeley.edu/~dmolnar/library.pdf.

177. National Business Review. (2004, July 8). Japanese children to be RFID'd. Retrieved July 11, 2004 from http://www.nbr.co.nz/home/column_article.asp?id=9531&cid=5&cname=Asia+%26+Pacific

178. R. Moroz Ltd. (2004, July). Understanding Radio Frequency Identification (RFID) (Passive RFID). Markham, Ontario: R. Moroz Ltd. Retrieved August 4, 2004 from http://www.rmoroz.com/rfid.html.

179. S.B. 1834. Radio frequency identification systems. California Senate, 2003-2004 Regular Session. Retrieved August 4, 2004 from http://www.leginfo.ca.gov/cgi-bin/postquery?bill_number=sb_1834&sess=CUR&house=B&author=bowen

180. San Francisco Public Library. (2004, March 31). RFID implementation: Proposed plan of action. Retrieved August 5, 2004 from http://sfpl4.sfpl.org/librarylocations/libcomm/rfidplan040104.htm

181. San Francisco Public Library. (2004, April 30). Privacy Policy. Retrieved August 5, 2004 from http://www.sfpl.org/librarylocations/libcomm/pdfs/privacypolicy.pdf

182. Santa Clara City Library. (2004, January). Checkpoint RFID and the Santa Clara City Library. Retrieved August 5, 2004 from http://www.library.ci.santa-clara.ca.us/rfid/checkpoint.html.

183. Sarma, E.S., Weis, S.A., Engels. D.W. (2002, November). White paper: RFID systems, security & privacy implications. Cambridge, MA: Massachusetts Institute of Technology, AUTO-ID Center.

184. Schneider, K. (2003). RFID and libraries: Both sides of the chip. Testimony presented at Committee on Energy and Utilities, California Senate, Nov. 20, 2003. Washington, DC. Retrieved from www.senate.ca.gov/ftp/SEN/COMMITTEE/STANDING/ENERGY/_home/11-20-03karen.pdf .

185. Smart, L.J. (n.d.). Considering RFID. Retrieved July 13, 2004 from www.csupomona.edu/~ljsmart/draftarticle.doc

186. Tien, L. (2004, January 10). RFID and libraries: EFF talking points for ALA IFC. Summary of presentation to the ALA Intellectual Freedom Committee of the American Library Association at ALA Mid-Winter, San Diego, California. Retrieved from http://www.privacyrights.org/ar/RFID-ALA.htm.

187. Want, R. (2004, January). RFID: A key to automating everything. Scientific American, 290(1), 56-66. Retrieved July 9, 2004 from MasterFILE Premier database.

188. Ward, D.M. (2004, March). March: RFID Systems. Computers in Libraries, 19-24.

189. Weiser, Mark. (1993, August 16). Ubiquitous Computing. Retrieved August 5, 2004 from

190. http://www.ubiq.com/hypertext/weiser/UbiHome.html 21

191. http://www.infosciencetoday.org/tag/rfid-for-library

192. http://www.infosciencetoday.org/tag/rfid-library

193. Dave Brich at Consultant Hyperion. URL:www.chyp.com . As quoted in "News Comment - Contact less crazy" in Journal: Card Technology; pub by Elsevier. p2, February 2005.

194. http://cplus.about.com/od/glossar1/g/xor.htm

195. Sudarshan S. Chawathe ,et al."Managing RFID Data" Proceedings of the 30th VLDB Conference, Toronto, Canada, 2004.

196. msdn.microsoft.com/en-us/library/aa933230(SQL.80).aspx

197. www.linfo.org/kill.html

198. http://www.inttelix.com/face-recognition-technology.php

199. Klaus Finkenzeller, "RFID Handbook, 2nd Edn" - translated by Rachel Warrington. John Wiley, 2003.

200. Steve Lewis, "A Basic introduction to RFID Technology and its use in the supply Chain". White paper by LARAN RFID Jan 2004.

201. London, United Kingdom, July 16, 2010 —(PR.com)—

202. Ashim A Patil, "i-TEK RFID Based Library Management Suite" a White paper. by Infotek Software & Systems PLtd, Pune, India. 02, April, 2004.

203. Lee Tien, Senior Staff Attorney. **EFF Talking Pointsfor ALA IFC.**Electronic Frontier Foundation http://www.eff.org/

204. Axiell http://www.axiell.co.uk

205. http://news.zdnet.co.uk/itmanagement/0,1000000308,2134438,00.htm

206. Bibliotheca http://www.bibliotheca-rfid.com/

207. "RFID for Libraries" at URL: http://www.biblio-tech.com/html/rfid.html.

208. "RFID Survival" at URL: http://www.rfidsurvival.com/contact.html

209. http://www.webopedia.com/TERM/B/bar_code.html

210. http://www.technovelgy.com/ct/Technology-Article.asp?ArtNum=58

211. http://www.technovelgy.com/ct/Technology-Article.asp?ArtNum=57

212. http://www.technovelgy.com/ct/Technology-Article.asp?ArtNum=20

213. http://www.wired.com/science/discoveries/news/2002/05/52343

214. Radio ID Tags: Beyond Bar Codes. Kendra Mayfield. 05.20.02 http://www.wired.com/services/feedback/letterstoeditor

215. http://www.autoidcenter.org/main.asp

216. http://www.wired.com/science/discoveries/news/2002/05/52343#ixzz0wgYcSocU

217. http://www.tutorial-reports.com/wireless/rfid/technology.php

218. Privacy. Debra Bowen, at a 2003 hearing. http://en.wikipedia.org/wiki/Radio-frequency_identification#cite_note-84

219. *Impact of RFID on Consumer Privacy.* Rich McIver, Special to RFID Gazette.

220. http://www.uc-council.org/

221. http://www.uc-council.org/global_tag/gtag_home.html

222. Page 9, RFID and Libraries, by Addison- Wesley/ Prentice Hall. Editors, Simson Garfinkel and Beth Rosenberg.

223. http://www.eff.org/Privacy/Surveillance/RFID/20031002_sfpl_comments.php

224. http://www.slais.ubc.ca/courses/libr500/04-05-wt2/www/T_Gnissios/privacy.htm

225. http://www.ala.org/Template.cfm?Section=ifresolutions&Template=/ContentManagement/ContentDisplay.cfm&ContentID=85331

226. Guidelines for Using RFID Tags in Ontario Public Libraries. http://www.ipc.on.ca/docs/rfid-lib.pdf

227. What Is NFC? http://java.sun.com/developer/technicalArticles/javame/nfc/

228. http://www.nfc-forum.org/home/

229. http://developer.gemalto.com/download.html

230. http://www.payezmobile.com/

231. http://touchandtravel.de/

232. NFC Technology: 6 Ways It Could Change Our Daily Lives May 6, 2010 by Sarah Kessler

233. http://mashable.com/2010/05/06/near-field-communication/

234. http://ttuki.vtt.fi/smarttouch/www/?info=whatissmarttouch

235. http://www.servtag.com/en

236. http://de.beta.friendticker.com/

237. http://mashable.com/category/facebook

238. http://www.nfriendconnector.net/

238. http://www.edi-center.com/edi-in-retail.htm

239. Mary Catherine O'Connor, "RFID - Based System

Tracked Victims of Hurricanes Gustav, Ike." RFID Journal, Nov.12, 2008

240. http://www.activewaveinc.com/applications_id_badges.php
241. RFID Competing Technologies http://www.tutorial-reports.com/wireless/rfid/competing.php?PHPSESSID=039f4561c2f386ac06a55515dac33e7a
242. http://www.activewaveinc.com/applications_overview.php
243. RFID Tag Characteristics. http://www.zebra.com/id/zebra/na/en/index/rfid/faqs/rfid_tag_characteristics.htm
244. Milan Korcok, "Tracking Itinerant Patients and Surgical Sponges," Canadian Medical Association Journal 180(8) (April 2009): E14-E15
245. IntiMetrix study, 2002 as quoted in "Lost Child Stats & Facts!" The Center to Prevent Lost Children; 2007, The Center to Prevent Lost Children, 21 April 2010 <http://www.wander-wear.com/pdf/StatsAndFacts.pdf>
246. Corie Lok, "Wrist Radio Tags." Technology Review, November, 2004, 25
247. Claire Swedberg, "World's Largest Cruise Ship Launches RFID-based Passenger - Tracking System." RFID Journal, February, 22,2010 <http://www.rfidjournal.com/article/view/7415>
248. Precision Dynamics Corporatist. Precision Dynamics Corporation's Smart Band® RFID Wristbands to Help Combat Counterfeiting at the South by Southwest (SXSW) Festivals. <http://www.pdcorp.com/en-us/company/pr2004-sxsw.html>. 09 Mar. 2004. Web. 22 Apr. 2010
249. http://www.activewaveinc.com/applications_inventory_control.php

250. http://nje-rfid.com/RFIDWIPTracking.htm
251. Junko Yoshida, "U.S. Navy uses RFID Technology to Track Wounded in Iraq—Radio - Frequency ID Chips in Wristbands Allow Doctors to Change or Create Records." Electronic Engineering Times, May, 26,2003, 16
252. http://www.activewaveinc.com/applications_manufacturing_line.php
253. Claire Swedberg, "RFID - enabled Device Sounds an Alarm When Someone Tries to Leave Home." RFID Journal, April, 7, 2010l
254. http://www.activewaveinc.com/applications_hospitals.php
255. Maria Kaganov. RFID: Document Tracking, Wednesday, September 1, 2010 - RFID Connections
256. http://www.zebra.com/id/zebra/na/en/index/rfid/faqs/rfid_basics.html
257. http://www.rcdtechnology.com/rfid-applications/success-stories/
258. http://www.engadget.com/2005/05/14/rfid-is-in-the-toilet/
259. http://www.usatoday.com/tech/news/techinnovations/2005-09-28-radio-chips_x.htm
260. http://news.bbc.co.uk/2/hi/technology/3697940.stm
261. http://www.vitacraft.com/rfiq/home.html
262. http://radargolf.com/home.asp
263. http://www.tutorial-reports.com/wireless/rfid/competing.php
264. http://www.informationweek.com/story/IWK20030110S0028
265. http://www.pg.com/en_US/gillette/index.jhtml
266. http://www.engadget.com/2006/03/13/us-issues-first-rfid-passports/

267. http://www.rfidgazette.org/2007/02/rfid_chips_go_t.html

268. India to adopt passive RFID technology for toll collection, Debdeep Chakraborty Wednesday, July 14, 2010, 11:18 Hrs [IST] http://www.projectsmonitor.com/ArticleList.aspx?sid=2

269. http://www.tutorialreports.com/wireless/rfid/environment/groups.php

270. http://www.fibre2fashion.com/industry-article/technology-industry-article/

271. Its boom time for RFID in India. http://www.rfid-weblog.com/50226711/its_boom_time_for_rfid_in_india.php

272. http://www.nol.ae/

273. http://www.acq.osd.mil/dpap/dars/dfars/html/current/252211.htm

274. http://www.epcglobalinc.org/standards/

275. http://www.ti.com/rfid/shtml/apps.shtml#overview#overview

276. http://www.tutorial-reports.com/wireless/rfid/environment/challenges.php

277. Das Internet der Dinge : Ubiquitous Computing und RFID in der PraxisVisionen, Technologien, Anwendungen, Handlungsanleitungen by Elgar Fleisch (Editor), Friedemann Mattern (Editor), Springer; (April, 2005)

278. RFID Handbook : Fundamentals and Applications in Contactless Smart Cards and Identification, by Klaus Finkenzeller, John Wiley & Sons (May 9, 2003)

279. RFID Labeling: Smart Labeling Concepts & Applications for the Consumer Packaged Goods Supply Chain, by Robert Kleist et al., Banta Book Group (August, 2004)

280. RFID in the Enterprise, by Tom Polizzi, WCCN

Publishing, Inc. (2004)

281. RFID Field Guide : Deploying Radio Frequency Identification Systems, by Manish Bhuptani, Shahram Moradpour, Prentice Hall PTR (February 10, 2005)
282. The Electronic Product Code (EPC): Naming Scheme for Physical Objects, by David L. Brock, MIT White Paper (2001)
283. Hardware Certification Paper, by EPCglobal Inc., (August 27, 2003)
284. http://www.tutorial-reports.com/wireless/rfid/environment/environment.php
285. http://www.dti.gov.uk/sustainability/weee/
286. http://www.gao.gov/new.items/d05849t.pdf
287. http://www.rfidjournal.com/article/articleview/780/1/1/
288. http://www.treehugger.com/files/2005/09/ecycle_electron.php
289. http://www.keepmedia.com/pubs/SupplyChainManagementReview/2004/09/01/640887/
290. http://www.mit.gov.in/content/radio-frequency-identification-rfid
291. tarmak007.blogspot.com/.../rfid-based-smart-cards-for-armed-forces.html
292. http://www.aviationweek.com/aw/generic/story_channel.jsp?channel=AviationWeek.com&id=news/awx/2011/03/29/awx_03_29_2011_p0 303166.xml&headline=India%20Explores% 20RFID%20Smart%20Cards%20For%20Military%20Personnel
293. Rafael Kleiman, and Rafik Loutfy.
294. Canada-India RFID project looks to improve traffic flow, reduce pollution. RFID technology may provide the key to better traffic management and improved pollution control in Canada, India and worldwide. June 19, 2008. http://www.physorg.com/

news133112758.html.

295. http://www.tutorial-reports.com/wireless/rfid/environment/csr.php

296. Ranjan Chaudhuri, Ravi Shankar, (2007) "RFID in retail industry: integration to application in Indian perspective", Journal of Advances in Management Research, Vol. 4 Iss: 2, pp.86 - 98

297. Indian Store to Demonstrate Retail Uses for RFID http://www.rfid-weblog.com/50226711/indian_store_to_demonstrate_retail_uses_for_rfid.php

298. http://www.creative-weblogging.de/cgi-bin/frames.cgi?url=http://www.wipro.com/newsroom/newsitem/newstory337.htm

299. Ravi Mathur, LOGISTICS: RFID - transforming the supply chain in retail. http://www.imagesretail.com/logistics_jun05.htm—

300. Kumar Dawada. Combating retail shrinkage with RFID. http://www.networkmagazineindia.com/200601/newsanalysis05.shtml

301. Rajesh Kumar RFID's Future Looks Bright for Indian Jewelers. http://www.rfidjournal.com/article/view/7288

302. RFID In Indian Market. http://www.oppapers.com/essays/Rfid-Indian-Market/87001

303. Samir K. Srivastava. Case Study: Introduction of RFID in the Indian Retail Market RFID and the Indian Retail market. Indian Institute of Management - Lucknow *Connective Technologies in the Supply Chain, Sameer Kumar, ed., Chapter 9, pp. 185-210, Taylor and Francis, 2007.*

304. Reliance Retail's RFID Plans. http://www.retailtechblog.com/2007/12/reliance-retail-rfid-plans.html

305. Expert view: Retail RFID today. http://indiaretailtimes.blogspot.com/2009/10/expert-

view-retail-rfid-today.html

306. Sharon Lobo, Oct 28, 2010 06:50 PM, Usage of RFID in retail is still not economically viable" http://www.cxotoday.com/story/usage-of-rfid-in-retail-still-does-not-have-an-economy-model/